美国流芳園设计

海外中国名园

〔美〕陈劲 著

Garden of Flowing Fragrance

Designing The Huntington's Chinese Garden

JIN CHEN

上海人民出版社

本书献给我最心爱的儿子陈汉廷

他是我的灵感源泉

I dedicate this book to my most beloved son Kevin H. Chen

who is my inspiration

目录 CONTENTS

序
PREFACE

写在《流芳园》出版之际

朱良志
Zhu Liangzhi

我在美国纽约大都会博物馆亚洲部做研究期间，办公室楼下就是中国展览区，展区并不大，在如此珍贵而狭小的区域里，却安排建造了一座室内中式园林，这就是著名的明轩。读书稍倦，我常到楼下回廊小憩片刻，目对园景，身与优游，陶然忘怀其间。园有中庭，四面以粉墙黛瓦与外相隔，院中靠墙置一半亭，傍以湖石假山，一丛芭蕉就在白色的墙前摇曳。偶尔听到潺湲的流水声，漏窗里忽有摇曳的竹影闪过，正陶醉间，一缕寒梅香气正入鼻中，此情此景，真使人有凌虚入空之感。

When I was doing research at the Department of Asian Art in the Metropolitan Museum of Art in New York City, there was an exhibition area of Chinese art downstairs. The area was not big, but there was an indoor Chinese garden court sitting within the limited space, which is the famous Ming Xuan or the Astor Court. Whenever feeling weary, I went downstairs to have a rest in the corridor of the garden. Looking at the scenes and being relaxed in the garden, I felt as though I disappeared among the scenes. The courtyard is surrounded with white-washed walls which separate it from the other part of the floor. There is a half-pavilion leaning against the wall, a couple of Taihu rocks, and a clump of plantains waving at the front of the wall. Occasionally I heard the sound of flowing water, and saw the shadow of bamboo swaying in the lattice window. When I was reveling in the scenes, a fragrance of wintersweet was rushing into my nostrils. In these circumstances, I felt as if I was floating high to the sky.

Ming Xuan is simple and elegant, tranquil and secluded. It narrates romantically the spirit of Chinese art. The court seems desolated, not dazzling, but is not shabby at all even though sitting with in the assemblage of the arts around the world in the Metropolitan Museum. The characters of *Ming Xuan* were selected from Wen Zhengming's calligraphy, which are delicate and vigorous. The exquisite design was done by Chen Congzhou, the most prestigious Chinese garden artist in the twentieth century of China.

Chen Congzhou was a garden master and a famous connoisseur of Chinese gardens. When

这明轩，朴素中有精致，简澹中有幽深，浪漫地书写着中国艺术的无上神采。她本是环睹萧然，远非琳琅罗列，放在大都会这一集人类艺术大成的海洋内，竟毫无愧色。“明轩”二字，集文徵明之书，风清而骨峻。玲珑的设计，出自20世纪以来中国最负盛名的园林艺术家陈从周先生之手。

陈先生是一位造园家，又是一位著名的园林鉴赏家。我在20世纪80年代开始步入美学与艺术研究之途时，对我影响最大的学者有两位，一位是宗白华先生，另一位就是陈从周先生。至今我还能记得当初读《说园》时的感受，真能以如饮慧泉来形容，我是由园林艺术走入研究中国艺术的世界的。坐于明轩之中，品读中国园林乃至中国艺术的内在韵味，更加理解大都会博物馆建造这座别致的中式小园的特别用心了。

近日，收到一部《流芳园》书稿，作者陈劲先生，是毕业于美国麻省的美籍中国著名园林建筑家，他是陈从周先生的学生，1983年毕业于同济大学建筑系，得到陈先生的亲自指导，曾参与过陈从周先生主持的豫园东部修复工程。流芳园是洛杉矶汉庭顿花园中的一座中国园林，我在美国期间，正逢此园部分开放，得以了解这座园林的具体风貌，感叹在此异国他乡的炎热之区，竟然有这样一座中国风韵的园林。从颇显方正的题有隶书“流芳园”匾额的正门步入，辗转流连，徘徊绸缪，她的厅堂楼阁，云墙篱落，亭桥假山，流泉涧瀑，

I started my career in the study of aesthetics and arts in the 1980s, the two scholars who had the most influences on my pursuit were Zong Baihua and Chen Congzhou. Today I still remember my feelings of the moment when I was reading the book of *On Chinese Gardens*, it was like drinking a spring of wisdom. I started my study on the Chinese arts beginning with the art of Chinese garden. Sitting in *Ming Xuan*, savoring the inner charm of Chinese garden as well as Chinese arts, I could understand the peculiar intentions of building this unique Chinese style courtyard by the Metropolitan Museum.

Recently, I have received a manuscript of the book *Liu Fang Yuan*. The author is Chen Jin (Jin Chen), who is a well-known American Chinese garden architect graduated from the University of Massachusetts. He was a student of Chen Congzhou at Tongji University, where he graduated in 1983 in China. He obtained Chen Congzhou's direct teaching and guidance. He also participated in the restoration project of Yu Garden which was directed by Chen Congzhou. Liu Fang Yuan is a Chinese garden located at the Huntington in Los Angeles. When I was in the United States, the garden was partially opened to the public. I visited the garden to my surprise to see this Chinese garden built in such a hot climate in a foreign country. Entering the garden through the front gate on which the quite formal *li shu* style of Chinese characters of "Liu Fang Yuan" inscribed in a brick horizontal board, I started to wander between the courtyards and pavilions. There are halls and

以至蕉情竹意，漏窗引景，曲径通幽之处，豁然开朗之时，都是活脱脱的中国风味。我猜测这一定是一位深通中式园林艺术的高手所构，甚至怀疑出自陈从周先生之手。

今得陈劲先生所赐书稿，方知此园营建之原委。中国传统园林家，有的长于营建，有的长于品鉴。兼营建与品鉴二者之长者，并不多见。像张南垣、石涛、戈裕良诸家虽有大制作，但他们品园之论并不多。明末计成是兼二者之长的艺术家，一生造园无数，又有《园冶》一书流传，说园林营建之种种。近世以来，当推陈从周先生，他园造得好，也说得好。接续此风者，陈劲先生最当其人，他不仅是一位造园家，又是一位卓越的园林理论家。《流芳园》一书，敷陈他营建此园的构思脉络，又体现出他对中国传统文人园林思想的深邃见解。

流芳园的格局颇大，是一座真正意义上的东方园林。在布局谋篇上对设计者有很高的要求。流芳园占地 75 亩，比苏州拙政园的规模还要大。坐北朝南，背山面水，建筑群多在北面，符合中式园林的堪舆之理。园中设小园九处，分别为春夏秋冬四时园和松涛园、幽竹园、盆景园、宝塔园和峪园。九园中各有其点景，以成“九园十八景”之体式。游园者由春园进入，依中国古代四时合四方的模式，春园在东，其中一座湖石假山停云峰格外引人注目。沿春园之回廊，进入夏园，但见园中湖光山色，汪洋一片，荷风澹荡，一座灰白色的石孔桥飞驾，临水构一四面厅，允为此园之要景。秋园取秋水寒山之意，置大片假山群，湖岸有画舫，

pavilions, Cloud Walls, bridges and rockery hillocks, streams and waterfalls, and scenery of plantain and bamboo, lattice windows leading to scenes behind, deep and secluded, open and spacious, all these are in the real Chinese style. Then I was guessing that it must had been designed by someone who was truly proficient in Chinese gardens. I had even thought that this garden might be designed by Chen Congzhou himself.

After received the manuscript from Jin Chen, I know now the whole story about the design and construction of this garden. Among the traditional Chinese garden masters, some were good at constructing gardens, while others were expert in connoisseurship. There were only a few people in the past who were skilled in both fields. For instance, though Zhang Nanyuan, Shi Tao and Ge Yuliang had done significant garden projects, they had seldom writings on the gardens. Ji Cheng of later Ming dynasty was an expert in both fields. He had done numerous gardens and handed down the book of *Yuan Ye* or the *Craft of Gardens* which covered almost every aspect of garden design and construction. In the modern times, Chen Congzhou was certainly the one who had done wonderful gardens as well as great writings on Chinese gardens. The one who has succeeded this practice is Jin Chen. He is not only a Chinese garden expert, but also an outstanding garden theorist. In the book of *Liu Fang Yuan*, Jin Chen has presented his train of thoughts on the design of the garden and his deep insights on the ideology of the traditional Chinese scholar gardens.

The layout of Liu Fang Yuan is quite large, which in a true sense is an authentic oriental

流芳园 - 入口门额砖雕
Liu Fang Yuan (LFY) - Garden Name brick carving

以供远眺。而冬园有藏书楼在焉。陈从周先生曾说，小园以静观为主动观为辅，大园以动观为主静观为辅，此当是其静观处。又有松涛园，取美国西部的松涛阵阵，来合东方之和鸣。精心营构的竹园，似有扬州个园之色彩。建在悬崖溪石上的峪园，很好地利用这里的地理特点，最得山林野逸之趣。而其中的盆景园，展示中国南北各派案上之佳作，方寸中见宇宙。流芳园中还有一座宝塔园，在古代园林构建中，平旷之景易成，高耸之塔难建，往往规格较高的园方有此式。而一般园林多是借景而成，如无锡寄畅园就是借周边佛塔而成。流芳园不避繁难，在园中垒起数丈高塔，成为全园中心，一个高潮点，一个登高望远的地方。全园多平面建筑，高不过两层。此塔一成，不仅成湖光塔影之制式，又使全园回环豫如之节律得以圆成。

garden. It requires a highly skilled designer to make a proper master plan and layout of the garden.The area of the garden is about 75 mu or 12 acres. It is larger than the area of the Humble Administrator's Garden in Suzhou. The layout of Liu Fang Yuan is oriented in north and south, fronting water and with mountains on the back, and most buildings and courtyards located on the northern side of the garden. All these are in accordance with the geomantic principles in Chinese gardens. There are nine small gardens within the whole garden, which are named as Spring Garden, Summer Garden, Autumn Garden, Winter Garden, Pine Garden, Bamboo Garden, Penjing Garden, Pagoda Garden, and Ravine Garden. Each of the nine gardens has its own key scenery, which made up of a garden structure of the "Nine Gardens and Eighteen Views". Visitors enter the whole garden from the Spring Garden which situated on the east according to the Chinese tradition of four seasons related to four directions. Inside the Spring Garden, a Taihu rockery named "Stopping Clouds" is particularly noticeable. Walking along the corridor, one arrives at the Summer Garden, where the visitor sees a panoramic view of the lake and mountains, the horizon is like an ocean, rippling lotus with fragrance, light gray stone bridges across the waters and a four-sided hall sits by the water which is the main scenery of the garden. The Autumn Garden, adopting the conception of autumn water and cool mountains, consists of a larger rockery hillock and a boat-shaped pavilion for looking out vistas. The Winter Garden has a library building and courtyard. Chen Congzhou once said: "In small gardens, 'in-position-viewing' should be predominant and 'strolling-viewing'

在因地制宜方面，流芳园有两点给我留下很深印象。中式园林是叠山理水的艺术，相地观势察脉，无水则脉络何成？山随水而活，桥缘水而成，阁在溪上，驳岸在水边，塔影映于湖，没有水，这篇文章就做不成。中国文人园林多在江南，当初乾隆复制苏杭园林于北京，造园家就感到水是一个难题。何况在南加州的热带之地，沙漠多，水难寻，此更不易做成。而最终园成，水体竟然成了园中主景，这不得不使人钦佩汉庭顿园主对建一座真正的中式园林的认真，也可以感受到设计者为之付出的智慧和艰辛。另外就是花木的配置。中式园林听香看舞，无花木之相配，纵然是华楼丽阁，终也梦难成。陈劲先生巧妙地利用南加州的热带植物，做成中国的诗情画意。如中国园林重古拙萧瑟的境界，老树高柳，盘旋映带，颇见趣味。陈先生利用南加州的橡树，创造东方美学的古朴之境，竟然天衣无缝，大解人颐。

流芳园的创造者并非简单地移植一个东方园林到此地。若考虑到流芳园坐落之所，更可明了其面临的挑战。第一，此园是巨大的汉庭顿花园中的一个园区。汉庭顿花园是举世闻名的人文之乡，是世界艺术和图书文献的积聚地，一个巨型的人文博物馆。在这样的天地中建一座中国的小园，你必须考虑到能立得起，不露寒蹇之相；而且要融进去，融入这篇人文的大叙述中。不是凭外在的大制作，需凭内在的艺术品质。第二，汉庭顿花园的主体是一个巨型的植物园，创造此园的汉庭顿先生平生最喜植物，这是一个植物的海洋，如何在此植物园林中造一座中国花园，显非易事。第三，汉庭顿内有沙漠园、日本园、莎士

supplementary; and vice versa." The Winter Garden is a place for "in-position-viewing". Nearby is the Pines Garden, where the American pine trees whisper in the wind, and echo the soughing of pines from the orient. The delicate Bamboo Garden reflects the charm of Ge Yuan in Yangzhou. The Ravine Garden is built in a gully by a stream, which well fitted into the terrain and was enchanting with wildness of wooded hills. The Penjing Garden is a different kind of world which displays a variety of best penjings with different styles of southern and northern China, which presents a cosmos through inches-sized landscapes of penjings. There is a proposed Pagoda Garden inside Liu Fang Yuan. In the history of Chinese gardens, it was easier to make a garden with sceneries expanding in horizontal directions, but difficult with a tall pagoda which was usually built only in higher standard gardens. Most gardens borrowed the view of a pagoda from outside the garden, for example, Ji Chang Garden in Wuxi borrows a view of a Buddha pagoda nearby. In Liu Fang Yuan, the design had not tried to avoid the challenge, instead proposed a tall pagoda to create a focal point, an apex and a spot to have panoramic views of the whole garden. Most of the buildings inside Liu Fang Yuan are low and flat, no more than two stories. Once this pagoda is built, there will be not only a dramatic scene of the lake and pagoda, but also a perfect finishing and essential touch for the ensemble of the whole garden.

With respect to the fitting in with the site, two aspects of Liu Fang Yuan have impressed

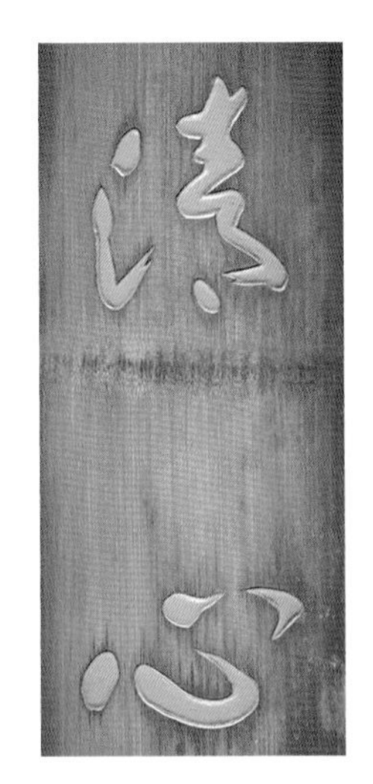

流芳园—“清心”对联局部
LFY - “Clear Mind”

比亚园、玫瑰园、亚热带丛林园等十多座花园。造流芳园之初衷，是为了提供一个了解中式园林的窗口，一个中西园林比较的空间。其中的日本园在 20 世纪初就建成，已有广泛影响。在西方思想的视域中，日本园林是东方园林的代表，那么，同属于东方、又是后建而成的流芳园如何确立自己的身份特点，这不能不说是造园者所面临的考验。

流芳园之初成，可以说承受住此挑战。其成功之处在我看来，正是对中国艺术内在精神出神入化地呈现。明轩和流芳园，师生二人之作，一在美国东海岸，一在西海岸，无声地诉说着东方艺术的韵味，说明东方文明所追求的美感。以小见大，见微知著，糜费无多，而佳景俨然在目。流芳园之所成，不光在目观，更在神游。天有时，地有气，材有美，工有巧，创造者妙心剔发，精心营构，以成一篇显现生命趣味之雅章。在这文化之府中造一文人小园，不靠炫耀知识，靠的是彰显中国文人意识中那种仰观宇宙、吐纳大荒的精神气质，靠的是敷衍韵人纵目、云客宅心的造园宗旨。老子说：“为腹不为目。”作为东方园林代表的中式园林主要是对心的，而不光是对目的。到此园中，不是看水，看山，水就那一汪，山就那一撮，并不是真正目的，而是看心，看生命趣味的流动，艺术家营造的是一个可与优游的世界，一个可与之气息相求的宇宙。

相比毗邻的日本园而言，流芳园也显示出她独具的特点。日式庭园以禅宗思想为底蕴，深染唐风，追求朴素，重视冥思，创造静寂的空间。而同样文化思想背景中的

me most. The Chinese garden is an art of arranging hills and waters. It involves surveying and observing in order to find out the characters and context of the site. If there is no water, how can the context be vivid? Mountains are alive because of waters; bridges have functions because of waters; pavilions stand beside waters; banks connect to waters; reflection of the pagoda sits in waters. So without waters, the “article” of making a garden would be unsuccessful. Most scholar gardens in China are located in the Jiangnan region. In Qianlong era, the emperor wanted to make copies of the gardens of Suzhou and Hangzhou in Beijing area, but the gardeners found that a lack of water was a difficult issue to resolve. Not to mention that in Southern California, where the weather is hot and dry, the land is mostly like a desert and water is difficult to find. And it is even more difficult to make a garden with waters. Liu Fang Yuan, however, actually becomes a water-centered garden. I had to admire the people at the Huntington for their seriousness to make an authentic Chinese garden over there. I could also imagine that the designer must had made a tremendous effort with a great talent during the design process. Another aspect is the plants in the garden. Inside a Chinese garden, listening to the fragrances, watching performances, all these dreams would not be realized if there are no plants in the garden, no matter how beautiful the halls and pavilions are. Jin Chen had skillfully integrated the tropical plants in Southern California into the Chinese poetic and picturesque compositions of the garden. In Chinese gardens, for instance, the beauty of

中式园林，由于民族习性有别，地理环境差异，历史承传中的脉络不同，宋元以来日渐形成她卓异的气质。中式园林是入世的，活泼的，非概念的，生生不已的哲学是中式园林的灵魂。中国园林追求涵纳深厚，静水深流，并不追求概念传达和冥思。即使是一个小园，也定然有通天之地，像苏州网师园、艺圃这样的小空间，也强调月到风来，揽八方之风物，收高天之云霓。

拿中式的假山与日本的枯山水相比，二者均非真山水，均是枯的。然枯山水妙在寂，假山妙在活。在中国艺术家看来，僵硬的石头中孕育着无限的生机；而在日本庭院艺术家看来，一片沙海，几块石头，就是寂寥的永恒。流芳园对此方面的处理，思路稳实，心意活络，在中式特色上做文章，主体建筑的安排，花木的配置，空间关系的处理，流动秩序的创造，乃至色彩关系的构建，一一得其时宜。这是非常不容易的。

我能够有机会流连在陈从周先生设计的明轩之中，今又得捧读陈先生法嗣陈劲先生的《流芳园》书稿，让我重新品味流芳园的实景，如此机缘，真似上苍对我这个园林爱好者的恩赐。记下我的粗浅感受，以就教于陈劲先生和诸同好。

2015 年 3 月 10 日于北京大学燕南园

quaintness and desolateness, old trees and tall willows, lingering and reflecting, all these present unique attractions. Jin Chen had made use of the California oak trees to create the poetic scenery seamlessly consistent with Chinese aesthetics. I was so happy to see this in the garden.

The creator of Liu Fang Yuan had not simply made a copy of an oriental garden and put it to the site. If you think about the context of the garden, you would understand how big the challenges were. First, it is one garden among other gardens in the whole Huntington ground. The Huntington is a world-famous cultural institution, a place that accumulates the art works and books and manuscripts of the world, and a huge museum of humanities. To build a small Chinese garden within this magnificent setting, you have to make certain that the garden can stand out, not looking shabby, and at the same time, blend it with the context so that it will become an integral part of the overall narrative of the Huntington. To achieve this goal, it relies not on an outer presentation of large construction, but on the inner artistic quality. Secondly, most areas at the Huntington are occupied by gardens. The founder Mr. Huntington was a life-long lover of plants. Here is a sea of plants. Within the botanical gardens, it is not easy to create a Chinese garden. Thirdly, there are sixteen different gardens on the Huntington ground, such as the Desert Garden, Japanese Garden, Shakespeare Garden, Rose Garden and Tropical Garden. The original idea to build a Chinese

流芳园 － 石拱桥、水榭
LFY - Stone bridge and pavilion

Garden may be to provide a window for visitors to know and understand the Chinese Garden and to compare with other western style gardens. The Japanese Garden was built in early twentieth century and became well known to the public. In the eyes of the Westerners, the Japanese garden is the model of oriental gardens. Then, how to define Liu Fang Yuan, which belongs to the same category of oriental gardens and would be built much later, needless to say, is a quite test to the garden designer.

The initial phase of Liu Fang Yuan has been completed and withstood the challenges. The success, in my opinion, lies in the fact that the garden is a superb representation of the essential spirit of Chinese art. Ming Xuan or Astor Court and Liu Fang Yuan, one work by the mentor and the other by his student, one on the East Coast and the other on the West Coast, have been silently presenting the charm of the Eastern arts and conveying the aesthetics of the Eastern civilizations. Inside the garden, one sees the large from the small, looks at the micro and knows the cosmos. The garden is built with limited cost but created with beautiful scenes. Liu Fang Yuan is not only for seeing, but also for the spiritual experience. The Heaven holds the time, the earth has Qi or energy, the material has its beauty and the craft has its artistry. The creators of the garden have worked meticulously on the design and construction and as a result, the garden becomes a vivid and elegant manifesto of life. Inside this cultural institution, this small scholar garden is not a show off of knowledge, instead it presents the spiritual quality of the Chinese scholar, the spirit of being open-eyed, open-minded and open-hearted to the universe as well as the purpose of garden-making to invite friends and to reside one's soul in the garden. Lao Zi said: "For the stomach not the eye." The Chinese garden, which is the prototype of oriental gardens, is not only for seeing but also for the heart to feel it. Coming to Liu Fang Yuan, one looks for not the water or mountain because the water is just a puddle and the mountain is just a rockery, which is not the real goal. The goal instead is for the heart, the vitality of life, a carefree journey to the world and a universe with which you can breathe.

Compared with the nearby Japanese Garden, Liu Fang Yuan has its own distinct characters. Japanese gardens were based on the context of Zen and greatly influenced by the Chinese gardens of Tang dynasty. Their spaces are simple, tranquil and contemplative. With the similar cultural background, Chinese gardens had gradually become more distinct since Song and Yuan dynasties because of the differences in national customs, geological environment as well as the historical context. Chinese gardens are secular, vivid and non-conceptual. Vitality is the soul of Chinese garden. The Chinese garden seeks the profoundness, like the tranquil and deep waters. It is not for presenting concepts or meditation. Although a small garden, the Chinese garden is a place where one can connect with the universe. For instance, the compact spaces of Wang Shi Yuan or the Master of the Net Garden and Yi Pu Garden in Suzhou, the wind and the moon are evoked, all things around are seized and the clouds in the sky are captured.

Both the Chinese rockery and the Japanese rock garden are man-made and dry, but the Japanese rock garden is excellent in its stillness, while the Chinese rockery is wonderful in its vividness. To the Chinese artists, the stiff rocks embody the abundance of vitality, while to the Japanese garden artists, an area of sands with a few rocks, presents the eternal solitude. In this regard, the design of Liu Fang Yuan is the result of a solid thinking and active mind. It emphasized the distinct Chinese features on the layout of buildings, the arrangement of plants, the organization of spaces, the order of sequences, and even the color patterns, all of which were properly designed. This was not easy at all.

I had an opportunity to wander in Ming Xuan designed by Chen Congzhou. And now, I am reading the manuscript of *Liu Fang Yuan* written by Jin Chen, an orthodox inheritor of Chen Congzhou, as if I am savoring again the real scenes in Liu Fang Yuan. It seems to me, a garden lover, that such a lucky chance is bestowed upon me by Heaven. I am writing down my feelings and humble opinions, and hoping to learn from Jin Chen and other like-minded colleagues.

Zhu Liangzhi
Yan Nan Yuan, Beijing University
March 10, 2015

(Translated from Chinese to English by Jin Chen)

流芳园 － 方舟待渡
LFY - Boat-pavilion be crossing

烟波罩良辰　Misty waves envelop a fine moment of morning.

流芳园 – 中心湖景

Liu Fang Yuan (LFY) - View of the Central Lake

前言
FOREWORD

园林，或一勺之庭院，或十笏之小园，或百亩之别业，或万顷之宫苑，都是人心目中的人间天堂！园林，是人类所向往的安居之所和精神家园！中国园林，亭台楼阁、池山林泉、佳木繁花，供人行、观、游、居、养，以妙通造化，陶冶情操，安顿心灵。中国园林，顺天地之理，成生生不息。

在世界园林艺术发展历史中，中国园林艺术以其悠久历史、丰富内涵、独特风格脱颖而出。中国园林的形成可追根溯源到上古时期先民对自然的崇拜和由此产生的神话传说，以及道家“道法自然”的思想。比如，昆仑神山、蓬莱仙居、一池三岛，以及灵台、灵沼和灵囿，目的都在于构筑一个任人啸傲的“仙境”。源远流长的中国哲学和美学思想，使中国园林呈现出“象天法地”和“范山模水”的主题。中国园林真正成为一门艺术是从魏

A garden, no matter whether it is a scoop-size courtyard, or a ten-yard-wide garden, or a hundred-acre mountain retreat, or a ten-thousand-hectare imperial garden, is after all a “Heaven on earth” in people’s minds. A garden is a comfortable place and spiritual home to human being. The Chinese garden, where the pavilions, terraces and halls sit alongside the hills and ponds among woods, flowers and streams, provides a place for wandering, sight-seeing, living and retreating. It is a place for reflecting, refining and meditating a person’s spirit as well. The Chinese garden conforms to the natural courses and embodies the vitality of life.

The art of the Chinese garden, with a long and rich history and distinguished style, holds a special place in the history of gardens in the world. The origin of the Chinese garden came from the concept of natural worshipping and mythological tales in the ancient China, and from the doctrine of “following the law of nature” of the Daoism as well. Examples of the legends are “the Sacred Mount of Kunlun”, “the Immortal Dwelling”, and “A Immortal Lake and Three Islands” as well as “the Spiritual Garden”, etc. The garden construction in China became an art in Wei Jin period (AD 220—420) . Before then, from Shang and Zhou dynasties (1600—256 BC) to Qin and Han dynasties (221 BC—AD 220), gardens were imitations of natural mountains and waters. From Wei Jin period, the gardens were mostly designed by scholars and painters. Since then, the garden construction became expressions of scholars’ virtues, thoughts, self-consciousness and aesthetic conceptions. Those expressions presented the essential characters of “Wei Jin Feng Liu”

东园图　明　文徵明　故宫博物院藏

Wen Zhengming (1470 - 1559), The Eastern Garden, The Palace Museum

晋时期开始的。之前从商周到秦汉的园林以利用和模仿自然山林为主，而魏晋的园林则开始以人造写意山水为主，开始显现出人的自觉意识，人的存在价值以及人格美与自然美相融合的思想，这也是“魏晋风流”的核心所在。唐代文人王维的“辋川别业”开启了中国“文人园林”的先河，以“心源”之山水构建园林，展现了抒情写意、寄情于景和诗情画意的园林精神。宋代之后直到明清的园林创作更加追求“境由心造”的境界，通过片石勺水的林泉之乐，体悟生命之意义，抒发心性之真情，寻求心灵之慰藉。

集儒家始祖孔子思想的《论语・雍也》曰：“知者乐水，仁者乐山；知者动，仁者静；

or the distinguished and admirable Wei Jin qualities. The earliest and most famous of this kind of gardens was the Wang Chuan Villa designed and owned by Wang Wei (AD 701—761) a preeminent painter, poet, calligrapher and musician in Tang dynasty (AD 618—907). The Wang Chuan Villa had become the prototype of the so called "scholar garden" or "literati garden" whose focus was on the expression of emotions, thoughts and poetical charms of the garden. From Song dynasty (AD 701—761) till Ming and Qing dynasties (AD 1368—1911), the design and construction of a garden was even more an expression of the emotions and thoughts of the designer or owner of the garden, where by taking pleasure in a few rocks and a small pond, one was able to understand the meaning of life, to express true feelings, and to seek the comfort of soul.

Kong Zi (551—479 BC), the founder of Confucianism, said in *The Analects* of Confucius: "The wise delights in waters, and the benevolent delights in mountains." Lao Zi (571—471 BC?), the founder of Daoism, said in *The Virtues of Dao*: "Man follows the earth, the earth follows the Heaven, the Heaven follows Dao, and Dao follows nature." Zhuang Zi (369—286 BC) said: "Heaven and earth coexist with me, and all things are unified with me as a whole." He further generalized the essential spirit of nature and said: "The Heaven and the earth have the highest virtue, but they do not speak a single word. The four seasons occur in regular cycles, but they do not raise a single argument. All things in the world grow in their own courses, but they do not give

流芳园 － 粉墙、黛瓦、绿茵
LFY - White-washed-wall,brick tile and the green

知者乐，仁者寿。"道家始祖老子《道德经》曰："人法地，地法天，天法道，道法自然。"道家庄子《庄子·齐物论》曰："天地与我并生，万物与我为一。"《庄子·知北游》进一步将天地万物之精神概括为："天地有大美而不言，四时有明法而不议，万物有成理而不说。"儒家强调亲和自然，道家崇尚回归自然。中国古代哲人强调人的心性和个体精神要与自然山水合一，将山水视为至善至美的代表。这些思想促成了中国山水文化独树一帜，别开生面。山水诗、山水画和山水园林的形成和发展，也使山水文化成为一门独立而博大精深的艺术。中国古代的文人雅士钟情山水林泉之乐，借山水以抒发情怀和慰藉心灵。东晋书法家王羲之"寄畅山水"，政治家谢安"寄傲林丘"，田园诗人陶渊明"性本爱丘山"。

a single explanation." The Confucians emphasizes the importance of being intimate with nature, while the Daoism returning to nature. The Chinese philosophers believed that man's temperament and spirit should be unified with nature. They had seen the consummate beauty of the natural mountains and waters as a metaphor of human perfection. This ideology had affected the nature of Chinese garden throughout the history. In China, the development of the *shan shui hua* (mountain-and-water painting) or the landscape painting, *shan shui shi* (mountain-and-water poetry) or the landscape poetry, and *shan shui yuan lin* (mountain-and-water garden) or the landscape garden had made the *shan shui* culture or the landscape culture the mainstream in the Chinese arts. This culture has become richer, more profound and extensive since then. The earliest scholars and designers of Chinese gardens were the idyllic poet Tao Yuanming (AD 365—427) of Jin dynasty, poet Wang Wei of Middle Tang dynasty and poet Bai Juyi (AD 772—846) of later Tang dynasty. Tao had his field and estate in the countryside, Wang had Wang Chuan Villa by the Wang Chuan River, and Bai had a humble cottage in the mist of Lu Mount. All these were the earliest prototypes of scholar gardens in China. These gardens became the "Heaven on earth" in the scholars' minds.

Xing Si (AD 671—740), a Zen master of Tang dynasty, once described the three stages of meditation as this: "At the beginning of meditation, one sees the mountain as a mountain and water as water; at the stage of understanding, one sees the mountain not as a real mountain and water not real water; and then at the stage of comprehension, one sees the mountain still as a mountain

中国早期杰出的园林艺术家当推东晋诗人陶渊明、盛唐诗人王维和中唐诗人白居易。陶渊明有返璞归真的“归园田居”，王维有诗情画意的“辋川别业”，白居易有寄畅林泉的“庐山草堂”。这些都是中国文人山水园林之宗。中国园林范山模水，将山与水结合于一壶公天地，为知者所乐，为仁者所爱，令人闭门而隐，俯仰山林，养生冶性，行义求志，陶冶情操。中国园林造就了一个理想的家园，一座心中的天堂！

唐代禅宗大师青原行思提出过参禅的三重境界：“参禅之初，看山是山，看水是水；禅有悟时，看山不是山，看水不是水；禅中彻悟，看山仍然是山，看水仍然是水。”其实，中国山水文化的真谛也就在此！中国文化中的“山水”是被赋予了人文内涵或人格化的山水。比如，中国画中的山水，取材于自然界中的真山水，初看像真的山水，再看又不像真山水，再用“心”看时它又是真山水了。中国园林中的山水亦是如此：初观，叠山理水，外师造化，宛自天开；再观，人造假山池沼，又非自然山水；进而观之，这假山假水又何尝不是真山真水呢？

中国明代造园家计成在其著名的造园理论书籍《园冶》中道出了造园的总体原则和宗旨，即“园林巧于因借，精在体宜”和“虽由人作，宛自天开”。《园冶》一书是留存至今中国最早、最系统、最全面细致地论述造园思想和造园手法的著作。计成以其丰富的造园经验、哲理意蕴和文采神笔，以诗文般的语言论述了中国的造园之道。中国艺术追求诗

and water still as water.” The Chinese landscape culture has the same true essence as the Zen. In Chinese arts, for instance, the representations of mountains and waters or landscapes are not merely imitations of natural mountains and waters. They are personified and cultivated with human emotions, thoughts and spirits. For example, inside a Chinese garden, at the first glance, the hills and waters are natural and real because they were built with natural rocks and water. But when one looks at them again, they seem as if they were fake and strange comparing with the real natural settings. Then, after strolling in the garden, one feels that the setting of the rockery hill, water and stream as well as the plants look like the real nature after all.

Ji Cheng (AD 1582—?), a Chinese garden master and designer in Ming dynasty, had written a book on Chinese garden design and construction, *Yuan Ye* or the *Craft of Gardens*. In the book, Ji Cheng pointed out that the most important principles on garden-making were that “a garden’s setting should ingeniously follow and borrow from the existing scenery and lie of the site, and be refined to fit the scale of the site and suitable for each other” and “though man-made, it should look like something naturally created”. The book of *Yuan Ye*, written between 1631 and 1634, is the first and earliest existing book on Chinese garden design and construction in the Chinese history. It presented the principles and methods of garden construction. Ji Cheng, with his abundant experiences in designing and constructing gardens as well as aesthetic thoughts and literature skills,

情画意，《园冶》一书可以说就是一首造园长诗，一幅心中的园林长卷！中国当代园林艺术大师陈从周先生在《说园》中谈到中国园林艺术时说过，中国园林是“文人园”，中国园林之所以百看不厌因为中国园林“有文化、有历史”，“中国的园林，它的诗情画意的产生，是中国园林美的反映”。他又说：“造园一名构园，重在构字，含意至深。深在思致，妙在情趣。园中有景，景中有人，人与景合，景因人异。有此境界，方可悟构园神理。”

中国园林以其深厚的文化底蕴和独特的艺术魅力也对世界园林文化的发展产生过影响。首先，中国文化及园林艺术对亚洲国家特别是对日本的造园艺术产生过巨大的影响。自中国魏晋时期（公元 4 世纪）中国文献典籍传入日本，隋唐时期（公元 7~9 世纪）中国文化艺术（包括建筑和造园艺术）及生活方式传入日本，到宋明时期（公元 13~15 世纪）中国的禅宗文化和写意山水画艺术传入日本。中国园林文化也随之传入日本，促进了日本园林的形成、发展和演变，包括日本园林中的“池苑”、“寝殿庭院”、“枯山水”（又称“唐山水”，“石庭”）以及“茶庭”等。其次，中国园林艺术也对西方园林产生了一定的影响，特别是中国自然山水园林对英国 17 ~ 18 世纪的风景式园林产生过直接影响，今天英国伦敦郊外的丘园中还留存有中国园林建筑和宝塔建筑。进入 20 世纪末期，中国园林艺术又被欧美国家的学者和民众重新认识和欣赏，并在欧美兴起了建造中国园林的风尚。

wrote the book in a poetic style to illustrate the art-and-crafts of garden construction. *Yuan Ye* can be seen as a long poem on garden design and construction, and a long scroll of splendid garden scenery in his mind! Chen Congzhou (1918—2000), a modern time Chinese garden master, said in his book of *On Chinese Gardens* that the Chinese gardens have been called the “scholar garden” because they have an enriched culture and a long history for which people are intrigued with them. “To build a garden is also called to compose a garden.” Chen said. “It is important to think well and to make subtle temperament out of it. There are scenes in the garden, and people are in the scenes, and then the people and the scenery are unified together. So each scene would be different because different people see and feel differently. Once reaching this realm, one can then comprehend the magical truth of garden construction.”

The Chinese garden, with its rich culture and unique charm, has influenced the outcome of some other gardens around the world. For instance, the Chinese garden had influenced the development of Asian gardens, especially the Japanese garden. In Wei and Jin period (AD 265—420) of China, Chinese literature and documents were exported to Japan. In Sui and Tang period (AD 581—907), Chinese culture and arts including the architecture and garden art as well as the life style were brought to Japan by the Japanese envoys and students. In Song (AD 960—1127) and Ming (AD 1368—1644) period, Chinese Zen Buddhism and scholar painting were exported to Japan. All these especially the Chinese garden art had great influences in the development

第一座原汁原味的传统式中国园林庭院是建在美国纽约市的大都会博物馆里的“明轩”。该项目于1978年由吾师陈从周先生主持，由苏州园林公司于1982年建成并对公众开放。迄今为止，在中国之外兴建的中国园林已有数十座。这些中国园林已经成为世人认知中国文化和园林艺术的具体媒介和杰出代表，并且已经融入当地的文化生活之中。

20世纪80年代，位于美国加利福尼亚州洛杉矶地区的“汉庭顿图书馆、艺术馆和植物园”开始酝酿在其内部建造一座中国园林。经过二十余年的努力，目前这座中国园林——“流芳园”已经初步建成并对外开放了。作为这座中国园林的总设计师，本人有幸借此机会将中国园林艺术和造园手法付诸实践，创作出一座新的传统式中国园林，同时得以将中国园林文化艺术展示给世人。

唐代著名画家张璪在《绘境》一文中提出了“外师造化，中得心源”的艺术创作思想，对中国的艺术创作产生了深远的影响。明代计成以其《园冶》一书为后人理解、运用和创作中国传统园林提供了精到的论述和指南。当代中国园林第一人陈从周先生的园林理论文集《说园》进一步丰富和完善了中国园林艺术理论。然而，如何运用这些造园思想和法则去结合汉庭顿的特殊场地而建造一座能体现中国园林艺术的园林，却是一项极具挑战性的工作。中国造园讲究“有法无式”和“因地制宜”。本书记述了流芳园的主要设计思路和

流芳园 - 漏窗、游廊、湖石
LFY - Lattice window, covered walkway and Taihu rocks

and formation of Japanese gardens including the Japanese water garden, the Zen garden, the Dry landscape and the Tea garden, etc. In the Western world, on the other hand, the Chinese garden had influenced the 17th to 18th century English landscape gardens. Even to today, people can still find some of the Chinese garden features in the Kew Garden outside of London, including pavilions and a pagoda. In the later 20th century, the Chinese garden has become a popular cultural exchange medium to the Western world in European countries as well as in the United States of America. The first authentic Chinese garden courtyard that was constructed outside of China was the Astor Court or Ming Xuan inside the Metropolitan Museum of Art in New York City. This project was directed and designed by Chen Congzhou in 1978, and constructed by a Suzhou garden company. The Astor Court was opened to the public in 1982. Since then, more than a dozen of Chinese gardens were built in other countries. The Chinese garden has become a medium for the local people and visitors to understand and appreciate the Chinese art, culture and garden.

In 1980s, people at the Huntington Library-Art Collections-and-Botanical Gardens, which is located in the great Los Angeles area of California, had started to think of building a traditional style Chinese garden inside the institution. After more than twenty years and with great efforts of many people, a newly built traditional style Chinese garden—Liu Fang Yuan or the Garden of Flowing Fragrance was opened to the public in 2008. As the Chief Designer of the garden, I was honored to have the opportunity to design the garden with applications of the principles and

创作理念，以及融入其中的中国园林文化和美学思想，阐述了流芳园所体现的中国园林艺术的“气韵生动”、“诗情画意”和“宛自天开”的美学境界。

“读”诗画以促进造园，“读”园林以感悟中国文化之精神。中国园林集中体现了中国人的哲学思想、艺术情趣、生活方式以及精神向往。晋代诗人陶渊明有诗曰:“静念园林好，人间良可辞”。唐代诗人白居易有诗曰：“天供闲日月，人借好园林。”明代书画家文徵明有题画曰：“人与青山已有约，兴随流水去无穷。”中国园林有诗一般的境界，画一样的意境。一花一世界，一水一性情，一石一精神。山水之美、建筑之美、花木之美以及天时之美尽收于一园。荷香浮动、雨打芭蕉、月移花影、松风听涛、雾失楼台，都给人以美的享受、精神的洗涤和心灵的安顿。

本书以中英文写作而成，中文为主体，英文则以中文之要意而概括叙述。希望本书成为“流芳园”的一部游园序曲，一座园林的叙述，一次心灵的优游。

甲午寒冬月
冶园道人 陈劲
写于上海知然居

methodology in the art of Chinese garden and to create a new traditional style Chinese garden at the Huntington. Liu Fang Yuan is a show case to the local communities and visitors and a place where people can appreciate the culture and aesthetic charm of the art of Chinese garden.

Zhang Zao (AD ?—1093), a famous painter in Tang dynasty, established a doctrine on Chinese painting, and said: “To learn from nature, then to create with one’s own thinking and feeling from the heart.” In terms of garden design and construction, Ji Cheng’s book of *Yuan Ye* would give a good guidance on how to design a Chinese garden. Chen Congzhou’s book of *On Chinese Gardens* would also provide a good direction on this subject. How to apply these principles, however, on the Huntington’s specific site, to make a great Chinese garden was a big challenge to me. In the field of Chinese garden design and construction, there are doctrines such as: “there are no fixed rules for designing gardens” and “to comply with the condition of the site”, etc. Therefore, this book of *Liu Fang Yuan* is meant to describe the design process and design concepts in the Liu Fang Yuan project. At the same time, this book also illustrates how the Chinese aesthetic and garden design principles complied with the site and environment in order to make Liu Fang Yuan a vivid and poetic garden at the Huntington.

To “read” a Chinese poem and landscape painting is for advancing the making of a Chinese garden; and to “read” a Chinese garden is for comprehending the spirit of the Chinese culture. The

流芳园
Liu Fang Yuan

Chinese garden epitomizes the Chinese philosophy, aesthetic interest, life style and the spiritual intention. Tao Yuanming, a master poet in Jin dynasty, said: "To meditate on the wonderful things of a garden, one can decline all the prosperous things in the world." Bai Juyi, a poet in Tang dynasty, said: "Nature provides the leisurely sun and moon, and people make use of good gardens." Wen Zhengming (AD 1470—1559), a scholar in Ming dynasty, said: "Human being has engaged with the mountains, and the enjoyment will last forever like the flow of a stream." The Chinese garden is like the realm of a poem and a work of painting. To the Chinese, a flower is a cosmos; the water has a temper; a rock has a spirit. All these, the beauty of mountain and water, buildings, plants and the charm of seasonal changes, have made a Chinese garden an enchanting place. The flowing fragrance of lotus flowers, the rain dropping on the plantain leaves, the moving shadow of flowers as the moon rises, the wind whispering among the pine trees, pavilions disappearing in the mist, all these will give one the enjoyment of the beauty, an enlightenment of the spirit and a comfort of the soul.

This book is written in both Chinese and English. The Chinese text is the principal version. The English version is written based on the main concepts of the Chinese version, so it is not translated word by word from the Chinese text though in some occasions transliterations are provided. I hope that this book will become a prelude to the visitors of the garden to enjoy Liu Fang Yuan or the Garden of Flowing Fragrance, a narration of the garden and a spiritual journey with leisure.

Jin Chen

A Garden Daoist Wrote in Zhi Ran Ju, Shanghai, China December 2014

天供闲日月，人借好园林

Nature provides the leisurely sun and moon; and people make use of good gardens.

流芳园 – 曲桥芳湖、亭台楼阁

Liu Fang Yuan - Bridges, the lake and pavilions

第一篇

梦的开始

Chapter One

The Beginning of the Dream

汉庭顿
THE HUNTINGTON

在美国加利福尼亚州南部的大洛杉矶地区，有一座名叫圣马利诺的小城。她的名字取自意大利的一个小公国——圣马利诺共和国。1877 年，美国人詹姆斯·苏伯和他的妻子马莉亚·威尔森为了纪念他们儿时在马里兰州的圣马利诺庄园，将他们在南加州的庄园也命名为圣马利诺。

1892 年，当时美国西部的铁路大亨亨利·爱德华兹·汉庭顿（1850 ～ 1927）先生应邀到苏伯的圣马利诺庄园做客，这个地方给汉庭顿先生留下了深刻的印象。十年之后，他将这个庄园买了下来。随后，经过一个世纪的营建和发展，这里已经成为世界知名的从事研究和教育的文化艺术殿堂和植物世界——“汉庭顿图书馆、艺术馆和植物园”。

In the great Los Angeles area, California, the U.S.A., there is a small city called San Marino. The city was named after the tiny republic of San Marino between Forli and Pesaro in Italy. In 1877, an American James De Barth Shorb and his wife Maria de Jesus Wilson christened their ranch San Marino in memory of Shorb's boyhood home in Maryland.

In 1892, a railroad tycoon in the West of the United States, Mr. Henry Huntington was invited by the Shorbs to visit their San Marino Ranch. Mr. Huntington was impressed with the Shorbs' place. Then ten years later, Mr. Huntington bought the ranch as a possible California home. Since then, after a century of developing and constructing, the ranch has now become one of the world's best research and education institutions in liberal art and arboretums—the Huntington Library, Art Collections and Botanical Gardens.

Mr. Henry Edwards Huntington (1850 – 1927) was a legendary man in the history of the United States. He was born in 1850 in New York State. When he was 20, he left his hometown and went to New York City. One year later, he started to work for his uncle's railroad business. In 1892, H. E. Huntington moved to San Francisco, the west of the U.S., with his uncle to develop the railroad enterprises and real estate business. They were very successful in the businesses. In 1903, he purchased the six-hundred-acre large estate of San Marino Ranch and moved to Los Angeles area. After that, he and his second wife Arabella D. Huntington worked together to develop the Ranch. They had devoted a great deal of energy and wealth to building the library, collecting art works and

亨利·汉庭顿先生
Mr. Henry E. Huntington

阿瑞贝拉·汉庭顿女士
Ms. Arabella Huntington

汉庭顿 1921 年全景
Aerial view of the Huntington in1921

汉庭顿艺术馆
The Huntington Gallery

汉庭顿先生在美国历史上可称得上是一位传奇人物：他于 1850 年出生于纽约州；20 岁离开故土到纽约市谋生，随后到他叔叔的铁路企业中工作；1892 年他来到美国西部城市旧金山继续开拓铁路和地产业务，并取得了巨大成功；1903 年，他移居洛杉矶地区并且购置了占地 600 英亩的圣马利诺庄园。之后，他和他的第二任妻子阿瑞贝拉·汉庭顿一起经营这个庄园，并且在图书和艺术品的收藏以及各式花园的建设上面投入了大量的精力和财力。他们于 1919 年正式将这座庄园命名为“汉庭顿图书馆、艺术馆和植物园”（以下简称

developing various gardens. They had officially named the estate as “the Huntington Library, Art Collections and Botanical Gardens” in 1919. The institution was opened to the scholars and the public in 1928.

Mr. Huntington was interested in reading and book collecting nearly all his life. His dream was to build a world class research library. His collections were focused on English and American histories and literatures. During the early years of his collection, he even bought some entire libraries or book stores from England, and then shipped them to his library in San Marino. By doing this, Mr. Huntington had collected abundant books and made the library well known as the “library

"汉庭顿"），并于1928年将其正式对学者和公众开放。

汉庭顿先生自幼就对读书和图书具有浓厚的兴趣，收藏图书是他一生的爱好。他的理想是建立一座顶级的研究型图书馆，图书内容着重于英国和美国历史及文学方面。在当时，他甚至不惜高价将英国的部分小图书馆和书店整个地买下，然后将图书运回到位于圣马利诺的汉庭顿图书馆中。因此，在短短几年之内就建成了一座内容丰富的优秀图书馆，并以"图书馆中的图书馆"闻名遐迩。

汉庭顿夫妇在艺术品收藏方面的成就也非常卓著。他们的收藏重点是18世纪的英国和法国艺术品，以及19世纪和20世纪初的美国艺术家作品，其中最突出和最有价值的是一些18世纪晚期的英国肖像画。除了对历史文献、图书和艺术品的兴趣之外，汉庭顿先生还特别喜欢园艺和植物。自从1903年汉庭顿先生购置下圣马利诺庄园之后直到今天，这里逐步建成了十六个不同风格的花园，收集了14000多个植物品种，其中最受人们喜爱的花园是以仙人掌科植物为主的沙漠花园，还有玫瑰园、日本园以及新近开放的中国园林——流芳园。

在流芳园建成开放之前，日本园是汉庭顿里唯一一座具有亚洲园林风格的花园。早在1911年年底，汉庭顿先生就从附近的帕萨迪娜市购置了一座日本茶室并将其移建到汉庭顿

of libraries".

Mr. and Mrs. Huntington were also excellent in collecting art works. Their collections included mainly the 18th century British and French art as well as the 19th and 20th century American art. Among the collections, the most famous and valuable paintings are some of the late18th century British portraits. In addition to collecting books, manuscripts and art works, Mr. Huntington was also interested in horticulture and gardens. From 1903 when he bought the San Marino Ranch till today, sixteen different gardens have been built at the Huntington with more than 14,000 different kinds of plants. Among the gardens, the most beloved ones are the Desert Garden with cactus plants, the Rose Garden, the Japanese Garden, and the newly opened Chinese Garden — "Liu Fang Yuan" or the "Garden of Flowing Fragrance".

Before Liu Fang Yuan, the Japanese Garden was the only one Asian style garden on the Huntington grounds. At the end of 1911, Mr. Huntington had purchased a Japanese tea house from the neighbor City of Pasadena and re-assembled it at the Huntington. He decided to put the Tea House in a valley on the western side of the property and to make a pond on a lower spot next to it. Later on, there was a bridge being constructed across the pond and more Asian plants being planted in this area. Although interested in the Asian gardens, Mr. and Mrs. Huntington didn't know very

里。他将这座茶室布置在一个山谷之中，随后在其周围建造了一座拱桥和一个池塘，并且种植了一些日本园林植物。在那时，虽然汉庭顿夫妇对东方园林十分好奇，很有兴趣，但是他们并不是很了解亚洲各国园林之间的风格差异。在随后的岁月里，汉庭顿的植物园工作人员在日本园里增添了更多的日本园林元素，直到 1957 年才正式将这座花园命名为“日本园”。

到了 20 世纪 80 年代，越来越多的人们开始意识到，中国是亚洲园林文化的发祥地，也是世界园林文化的先驱者。日本园林原本是从中国园林演变而来的，但二者在风格上有所不同。另外，就园林植物品种来讲，世界各地乃至北美洲地区，包括汉庭顿花园里的许多植物，都是源自中国的。所以，独立建造一座中国园林的梦想在汉庭顿内部开始形成，而且吸引了包括大洛杉矶地区多个社区的追梦者。

much about the very differences between the various gardens in Asian countries. As the time passed by, the gardeners started to put more Japanese garden elements and plants inside the Japanese Garden area. In 1957, the Huntington had officially named the garden as the “Japanese Garden”.

During the 1980s, more and more people at the Huntington had started to realize that the Chinese garden was the origin of the Asian gardens and represents one of the most ancient garden cultures in the world. The Japanese garden, in fact, evolved from the Chinese garden art in the early times. Later on the Japanese garden had developed into some distinguished features which became quite different from the Chinese garden. Furthermore, in the botanical world, there are many plants that came from China originally. For all these reasons, a dream to build a Chinese garden at the Huntington has been generated since then. The dream has attracted many followers within the Huntington institution as well as the local communities in the great Los Angles areas.

汉庭顿先生生前的官邸

The former Mansion of the Huntington

汉庭顿官邸平台

The Balcony of the Mansion

追梦人
THE DREAM PURSUERS

自1912年开始，汉庭顿夫妇对亚洲园林的兴趣和梦想一直伴随着整个汉庭顿的发展而增强。在建园最初的几年里，汉庭顿先生有幸找到了一位园林植物专家，德国裔美国人威廉·荷瑞奇，让他做植物园主任。荷瑞奇的工作内容之一就是带领着园艺工们对日本园进行了扩建和充实，包括添置了铜钟鼎、木拱桥以及水泥做成的假山等，使得日本园更具“东方气息”。在1928年，汉庭顿作为一个非营利性文化机构正式向公众开放。在随后的岁月里，越来越多的人来到汉庭顿参观和游览，同时也对日本及东方文化有了初步的了解和认识。

20世纪90年代，汉庭顿机构组建了一个“花园监管委员会”，由时任植物园主任詹姆斯·富尔森先生负责中国园的筹建工作。他的第一项工作是邀请了移居美国的原同济大学建筑系教授朱雅新女士，以及当地建筑师奥芬豪斯先生和景观设计师本内特先生一起，

Beginning in 1912, Mr. and Mrs. Huntington had the interest and dream to develop Asian style gardens on the Huntington grounds. The dream had become bigger and more real along with the overall development of the Huntington in the past years. During the early years of the development, Mr. Huntington had hired German-born William Hertrich as the director of the botanical gardens at the Huntington. Mr. Hertrich was an expert in horticulture and gardening. One of his first tasks at the Huntington was to develop the Japanese Garden by upgrading the existing facilities and adding more Japanese garden elements such as a bell and bell pavilion, a couple of stone lions, an arched bridge, and a cement rockery, etc. All these had brought the garden more “oriental appearance”. Since then more visitors have started to understand the Japanese Garden and some of the Asian cultures.

In the early 1980s, the Huntington had established a committee called the “Gardens Overseer Committee”. James Folsom, Director of the Botanical Gardens, was in charge of the initiative for the Chinese garden project. His first task was to invite Zhu Yaxin (Frances Tsu) , an architecture professor at Tongji University in China before immigrating to the U.S., Bob Ray Offenhauser, a local architect, and Todd Bennitt, a local landscape architect, to select a site and to produce a preliminary design for the Chinese garden. The preliminary design had helped the members in the Gardens Overseer Committee to understand some of the concepts and characteristics of Chinese gardens. One of the members, Mr. Peter Paanakker was particularly interested in the project.

流芳园 – 橡树与“爱莲榭”
LFY - Love for the Lotus Pavilion behind the oak trees

在汉庭顿内为中国园初步选择了一块用地，并提出了一份初步构想图，目的是使花园监管委员会的成员们可以理解中国园林的一些基本特征。其中一位委员皮特·潘纳克先生对建造中国园项目特别感兴趣并予以大力支持，他于 2000 年将其一千万美元的资金捐赠给汉庭顿作为中国园项目的基金。这项捐款使得中国园项目得以实质性地启动，并由富尔森主任负责组织和推进中国园项目。我是受富尔森主任邀请于 2000 年 11 月正式入职汉庭顿机构，并作为中国园的总设计师和项目经理开始了为期三年的规划设计和项目管理工作。汉庭顿中国园项目团队在现任汉庭顿总裁史蒂芬·郭必列博士的亲自领导下，一步一步地将建设中国园的梦想变成了现实。

In 2000, he had donated a $10 million to the Huntington to construct the Chinese garden. This remarkable gift had made a real start of the project. Since then, James Folsom has been responsible for the operation of the project. I was invited by James Folsom, and hired by the Huntington to work on the Chinese garden project as Chief Designer and Project Manager in November 2000. And then I had worked at the Huntington for three years on this project. Steve Koblik, President of the Huntington, has led the Chinese Garden project team to work unremittingly for years. All these great efforts have made the dream to build a Chinese garden a reality.

可以说，汉庭顿中国园（现称为“流芳园”）从梦想开始到今天的初步实现，是由于一群“追梦人”的思考、创新和努力才得以一步步地完成的。这些“追梦人”包括：汉庭顿夫妇、汉庭顿机构的历任总裁们、植物园主任们、花园监管委员会的委员们、中美双方的设计师们和建造者们。同时，也包括本项目的赞助者、当地的华人社会团体和个人，以及部分企业的支持者们。“流芳园”如今已经成为中国园林艺术及文化的传承者和传播者。

It is true that the realization of the dream of building a Chinese garden, now Liu Fang Yuan or the Garden of Flowing Fragrance at the Huntington, is the continuous and collaborative efforts from the “dream pursuers”. Their ideas, creativities and hard working have made the dream a reality. These “dream pursuers” include Mr. and Mrs. Huntington, the leaders and staff at the Huntington, the Gardens Overseers, the garden docents and volunteers at the Huntington, the design and construction teams from both China and the U.S. as well as the donors and local communities. Today, to the public, visitors and scholars, Liu Fang Yuan has become an inheritor and a medium of the Chinese culture and garden tradition.

流芳园 － 入胜

LFY - Entering a World of Attractions

流芳园 － 入口门楼
LFY - The Entry Gate

第二篇

人文天地

Chapter Two

The Scholars' World

人文环境
The Cultural Environment

自从汉庭顿先生于1903年购置了圣马利诺庄园之后，汉庭顿夫妇便将他们对人文、艺术和自然的极大兴趣和挚爱转化为对汉庭顿的营造。他们的睿智以及独特的鉴赏力使得汉庭顿图书馆、艺术馆和植物园成为一处人文气息十分浓郁、学术水平十分优秀、自然环境十分优美的世界知名文化机构。这里的人文内涵是从图书的收藏开始的。汉庭顿先生一生都对收藏书籍和文献具有浓厚的兴趣。他儿时的书籍至今还珍藏在这里的图书馆中，其中一本是《为在家里的小孩子的歌集》。

在1911年至1925年间，汉庭顿不但购置了他感兴趣的图书，而且更多地是将部分英国和美国的图书馆整个地买了下来，同时收集了大量的名家手稿、书信和图片等。这方面的收藏极其丰富且珍贵，比如这里拥有中世纪英国作家杰弗雷·乔叟所著的《坎特伯雷故事》

After the purchase of the San Marino Ranch in 1903, Mr. and Mrs. Huntington had devoted their great interests and love of the humanities, art and nature to the development of the Huntington institution. Their farsighted vision and unique appreciation had made the Huntington Library-Art Collections-Botanical Gardens one of the best and most famous institutions in the world. Here, the humanities are enriched, the academic level is the state-of-the-art, and the ground is very beautiful. This cultural environment has started with the book collecting. Mr. Huntington was interested in book and manuscript collecting nearly all of his life. One of his boyhood books, *Songs for the Little Ones at Home*, was still kept in the library.

Between 1911 and 1925, Mr. Huntington had not only collected books but acquired a number of very important libraries from Britain and America. At the same time, he had collected an enormous number of manuscripts and letters. Among the rare books in the collection, there were the Ellesmere manuscript of Geoffrey Chaucer's *Canterbury Tales*, the *Gutenberg Bible*, the first substantial book printed with movable type in Europe in 1455, the early Shakespeare editions in the Renaissance period, and the 18th century British novelist Charles Dickens' early editions. In American history and literature, the Huntington had collected the manuscripts and letters from George Washington, Benjamin Franklin and Abraham Lincoln as well as the novelist Mark Twain's manuscript of *The Prince and the Pauper*, etc. Recently, another unique collection which was found in 2014 is some of the volumes of the *Yong Le Da Dian* of Ming dynasty of China. This finding has

《坎特伯雷故事》

The Canterbury Tales

《古腾堡圣经》

The Gutenberg Bible

MR. WILLIAM
SHAKESPEARES
COMEDIES,
HISTORIES, &
TRAGEDIES.
Publiſhed according to the True Originall Copies.

LONDON
Printed by Iſaac Iaggard, and Ed. Blount. 1623.

莎士比亚剧本

The Shakespeare editions

图书馆藏书

The collections of books

件精美的雕塑等欧洲艺术品，以及近 300 件美国的绘画和雕塑作品。

在文献和图书收藏方面，迄今为止，汉庭顿图书馆已有超过七百万件名人的手稿或作品原稿，四十余万本珍贵而稀有的图书和资料，一百三十余万件照片、印刷品和缩微胶片。这里已经成为世界上最好的研究英美历史和文学的图书馆之一。现在，每年有近两千名来自世界各地的学者到汉庭顿来做研究、交流和教学工作，还有近五十万的参观游客。因此，汉庭顿可称得上是世界上最具魅力的研究人文历史和鉴赏艺术作品的环境之一。这里已经成为一处名副其实的“人文天地”！

who want to study the histories and literature, and to appreciate art works. Evidently, the Huntington is worthy of the name “Scholars’ World”.

《长腿》
The Long Leg

汉庭顿图书馆内景

The inside view of the Huntington Library

Executive Mansion
Washington, April 30. 1864

Lieutenant General Grant.

Not expecting to see you again before the Spring campaign opens, I wish to express, in this way, my entire satisfaction with what you have done up to this time, so far as I understand it. The particulars of your plans I neither know, or seek to know. You are vigilant and self-reliant; and, pleased with this, I wish not to obtrude any constraints or restraints upon you. While I am very anxious that any great disaster, or the capture of our men in great numbers, shall be avoided, I know these points are less likely to escape your attention than they would be mine– If there is anything wanting which is within my power to give, do not fail to let me know it.

And now with a brave Army, and a just cause, may God sustain you.

Yours very truly
A. Lincoln

林肯手稿

Abraham Lincoln' s letter

格林家具

The Greene & Greene furniture

花园世界
A World of Gardens

汉庭顿还有一个独特之处，即在如今拥有的207英亩的土地上，建成了十六座风格各异的园林或花园，花园的占地总面积达120英亩。这些花园不仅提供了人们游览、观赏和休憩的场所，而且以其不同的风格和植被，也反映出了不同的自然环境以及不同地域文化的多样性。

汉庭顿的植物园是在1903年汉庭顿先生购置的600英亩圣马利诺庄园的土地上逐渐形成的。最早，这里是一片以柠檬树和橙树为主的果园，随后逐步开始建造不同的花园。首先是百合花池塘、棕榈园和沙漠花园，之后是玫瑰园和日本园。到了1927年，汉庭顿庄园的总面积减少到两百英亩左右，其中一半的土地被改造成了花园。汉庭顿先生本人特别喜欢植物，并且喜欢看着它们从种子到长成的过程。他和植物园主任荷瑞奇一起从世界各地

The Huntington institution has another unique feature that, within a 207-acres property, there are sixteen different style gardens as today. The total area of these gardens covers about 120 acres. These gardens are not only places for visitors to enjoy, appreciate and relax, but also with their various plants and distinguished characters, representations of a variety of cultures and garden settings.

When Mr. Huntington purchased the San Marino Ranch in 1903, the total area of the property was 600 acres. At the beginning, most area of the land was covered by lemon trees and orange orchards. Later on, the gardeners at the Huntington had started to build one garden and another throughout the years. The first garden was the Lily Pond Garden, then the Palm Tree Garden, and then the Desert Garden, after that the Rose Garden and the Japanese Garden. By the year of 1927, the total area of the Huntington was reduced to 200 acres, where a half of the land was turned into the gardens. Mr. Huntington was very interested in horticulture and plants. He liked to see a plant growing from a seed to a bigger plant. He and William Hertrich collected a lot of seeds from various places in the world and planted them in different gardens. Therefore, there were many kinds of plants inside the gardens which made the Huntington a place of true botanical gardens. Among these gardens, the North Vista Garden is a very European style garden. It has an axle view, a big open lawn, a limestone fountain, and a row of plaster figure sculptures, etc. Today, there are in total about 14,000 different kinds of plants in the botanical gardens. The gardeners have planted

汉庭顿玫瑰园

The Rose Garden at the Huntington

收集了各种植物的种子，并将它们分门别类地种在不同的花园里，所以这里的植物品种十分丰富，最终形成了一个名副其实的植物园。在这些花园中，最具欧洲风情的花园是意大利风格的“北远景”景点。这里有空间对称的轴线，几何形的大草坪和石制喷水池，以及排列于草坪两旁的欧式风格的人物雕像。整个汉庭顿植物园内的植物品种超过了一万四千种。园丁们根据不同花园的特征和主题，将这些色彩缤纷、姿态俊逸的树木花草种植在各个花园内。这里既有茂密丛林，池塘叠水，疏林草地，又有沙漠奇木，四季花卉，奇花异果。这里已经成为植物学家和园艺爱好者以及孩子们的“乐园”，是一个非常理想的进行园艺植物科研和教学的场所。

the various plants according to the nature and characters of each garden. Here at the Huntington, there are various garden features such as small forests, ponds and water cascades, woods and lawns, as well as desert plants, seasonal flowers, and rare flowers and fruits, etc. The Huntington gardens have become a "paradise" to the visitors, plant lovers and children. It is an ideal place to do learning and researching on the horticulture and botany as well.

Mr. Huntington was a successful businessman, an art connoisseur and collector, a nature lover, and also a "scholar". We can say metaphorically that Mr. Huntington had integrated the literature, art and nature into "a great article" which was so vivid and rich. He had unselfishly given out his

汉庭顿先生是一位成功的商人，是一位有品位的鉴赏家和收藏家，是一位大自然的热爱者，他更是一位“文化人”！他把一篇融文学、艺术和自然为一体的“大文章”做得有声有色、丰富多彩，并无私地奉献给了世人。这篇“大文章”展示出了人与人、人与自然的和谐关系，它感化和启迪了来此观光、游览、学习和研究的人们。当人们走进这个人文天地和花园世界时，无不被这里的书香、花香和迷人的艺术品所感染、所触动。人们甚至会忘却日常的生活乃至名利的诉求，而进入到一个纯粹审美的境界之中。

“great article” to the world. This “great article” has presented the harmonious relationships between man and man, man and nature. It has inspired the visitors, researchers and students to appreciate, to learn and to communicate with the literature, art and nature. When people visit the library and art galleries or wander in the gardens, they are inspired and moved by the scent of books, the elegance of art works and the fragrance of flowers. They may enter a purely aesthetic state and forget the daily chores or the desires for the fame and fortune.

汉庭顿日本园
The Japanese Garden at the Huntington

汉庭顿沙漠园

The Desert Garden at the Huntington

曲院修竹里，蕉庭别洞天

Behind the slender bamboo,
the secluded plantain courtyard lies with unique beauty.

流芳园 – 入口庭院
Liu Fang Yuan – Entry Court

第三篇

场地精神

Chapter Three

The Spirit of the Site

相地合宜

THE PROPER SITUATING

汉庭顿是一个集人文、历史、艺术和造园为一体的世界著名的文化研究和教育机构。她将人文与自然结合得如此和谐与美丽，这在世界各地的文化机构和植物园中是很少见到的。要在这样一个环境中建造一座中国园林，首先面临着三个问题：一是建在一个什么样的场地环境中，二是这座园林的功能和风格定位是什么，三是如何将中国园林融入汉庭顿的人文和自然环境中。

20 世纪 90 年代中期，汉庭顿的团队在这方面已经进行了初步的尝试。植物园主任富尔森先生与建筑师朱雅新先生、建筑师奥芬豪斯先生以及景观设计师本内特先生一起在汉庭顿内初步选择了一块用地，构思出春、夏、秋、冬四个主题园，并勾画出了中国园的雏形。但在当时，这项工作还非常粗浅。从 2000 年 11 月正式到汉庭顿工作开始，作为中国园的

The Huntington is a world-renowned research and educational institution that assembles the literature, history, art and gardens. It has created a unique setting that integrates the humanity with nature in such a harmonious and beautiful way that one can hardly find the same kind of institutions around the world. In this unique context, to design and build a traditional style Chinese garden, one would have to deal with three fundamental issues at the beginning of the project: first, what kind of site would be the best for the garden; secondly, what the functions of the garden are; and thirdly, how the Chinese garden blend with the Huntington's cultural and natural environments.

In mid 1990s, the Huntington team had started to work on those issues. Director James Folsom, architect Frances Tsu, Bob Offenhauser and landscape architect Todd Bennitt had worked together to select a site and produced a preliminary design for the Chinese garden. The main concept was to create four small gardens surrounding a pond to represent the four seasonal features: spring, summer, autumn and winter. Although the design had shown some of the scope and characters of the garden, it was very preliminary and undefined. When I started to work at the Huntington in November 2000, my first task was to look attentively at the site and the surroundings. In the meantime, I started to review the preliminary design and to think about how to make the garden fit into the site in order to create an authentic Chinese garden here at the Huntington.

汉庭顿航拍图

The aerial photo of the Huntington

总设计师和项目经理，我首先对原来的选址场地进行了仔细勘察，同时对原有的雏形方案进行了重新思考，重点是根据场地的特质来设计出一座真正意义上的中国园林。

中国明代著名造园家计成于 1634 年完成的造园巨著《园冶》共三卷十四篇。计成在首篇《兴造论》中就提出："凡造作，必先相地立基"，"园林巧于因借，精在体宜"；在《园说》篇中提出："虽由人作，宛自天开"；在《相地》篇中提出："园基不拘方向，地势自有高低；涉门成趣，得景随形，或傍山林，欲通河沼"，"相地合宜，构园得体"；在《借景》篇中指出："构园无格，借景有因"。计成所阐述的造园思想和法则可以作为汉庭顿流芳

Ji Cheng, a garden master of late Ming dynasty in China, wrote the book of *Yuan Ye* or the *Craft of Gardens* in 1634. It contains three volumes and fourteen chapters. In the first chapter of "On Garden Construction", Ji Cheng wrote: "Whenever to make a garden, the first task is to study the site and make foundations." "A garden should follow and borrow from the existing scenery and lie of the land, and be suitable to the site with right proportions to each other." In the chapter of "On Gardens": "Though man-made, it should look like something naturally created." In the chapter of "Situating": "The foundation of a garden should not be restricted to any particular direction; the land will have its natural highs and lows. The gate to the garden should be attractive; the scenery

园的规划设计指导思想。第一项工作自然是“相地”，即场地勘察，或系统地“相地”、“观势”和“察脉”，亦同战国时期《考工记》中所说的“审曲面势”，即全面深入地了解场地的地形地貌及其所含的元素和特性。相地观势察脉的工作是设计一座园林关键的一步。

流芳园的最初选址是在一片低洼地上。周围有树林和灌木，平时这里不会积水，只有在雨季才会形成暂时的池塘。中国园林，无水不成园！鉴于水在中国园林中占有的重要位置，应将这片低洼地改造成一个水体，或池塘或湖泊，再环以园林建筑、庭院、假山和花木等，便可大致形成一个具有中国特色的园林格局。但是要真正达到“巧于因借，精在体宜”和“虽

follow the natural lie of the land, whether the garden lies beside wooded hills or abuts on a river or pond", and "a site should be chosen appropriately; the garden should be made appropriately". In the chapter of "Borrowing Sceneries": "There're no fixed rules for designing gardens but there are reasons to borrow sceneries." All these concepts and principles mentioned in Ji Cheng's book should be applied to the making of the Huntington's Chinese Garden. Therefore, the very first task should be the "situating" or "site investigation" for the garden, which includes surveying, mapping and investigating the site. "To look carefully on the surface and find the situation," said in the book of *Kao Gong Ji* or the *Record of All Crafts* (written in 400—300 BC). Observing and finding the

流芳园场地中心区
The central area of Liu Fang Yuan site

流芳园场地中心区
The central area of Liu Fang Yuan site

由人作，宛自天开”的园林意境和艺术效果，还必须对这片环境作细致深入的勘察，必须具体了解地形的起伏、地势的曲深变化、花木的品种和姿态、景观视线和借景的条件以及场地与周边环境的关系等，之后才可以把握这个场地的特质和灵气，从而设计出一座独具特色的中国园林。

这项“相地”的工作我前后进行了三个月的时间，从 2000 年 11 月开始直到 2001 年 2 月结束，也就是洛杉矶地区的冬季和春季。在这个过程中，我在整个场地里步行观测了几十次，了解并记录下每一个地形的标高，每一棵大树及每一丛灌木的品种和位置，水系去

characters and spirit of the site is a crucial step in the design of a garden.

The site initially selected for Liu Fang Yuan was at a lower spot on the western side of the Huntington ground. The central area of the site was usually flooded during a big rain, so it would become a pond or small lake then. The site is surrounded with trees and shrubs. For a typical Chinese garden, there should be water in it; otherwise, it won't be called a garden in the Chinese tradition. Because of the importance of water in a Chinese garden, the lower area of the site should be turned into a lake or pond, and then surrounded the lake with pavilions, courtyards, rockeries and plants. After doing all these, a typical Chinese garden would be formed. If the garden would be truly

向和曲折，土质的松软程度，主要视线的方向、角度和对景空间关系等。而且，我还特意观察了季节和气候变化在场地中的反映，以及每天不同时辰相对应的环境变化特点，包括朝阳、晚霞，甚至包括月亮升起的角度和时间等。经过这三个月的观察和勘测，可以说我对这片场地的特征已经“了如指掌”了！

constructed according to the principles of "following and borrowing from the existing scenery and lie of the land, and being suitable for the site with right proportions to each other" as well as "though man-made, it should look like something naturally created", a careful and in-depth investigation of the site must be done before designing the garden. The information of the site must be collected in terms of the highs and lows, the spatial characters, the kinds of the plants and their gestures, view corridors and sceneries to be borrowed from as well as the relationship between the site and the surround areas, etc. Once the essence and characters of the site or ground are comprehended, then it is possible to design a unique Chinese garden for this particular site.

流芳园场地南端

The southern side of Liu Fang Yuan site

流芳园东侧雪松林
The cedar trees on the eastern side of Liu Fang Yuan site

I had spent about three months on the "situating" task from November 2000 to February 2001, when it was the winter and spring times in Los Angeles area. During this process, I had walked on the site more than fifty times. I had mapped out the site and put marks of the elevations, contours, the locations of the big trees and shrubs, the running directions of the stream, the soil types, view corridors, view directions and angles, etc. I had even watched and recorded the schedules and directions of the sunrise and sunset as well as the moon appearing during the day and night. After the three months' investigation of the site and its surroundings, I could say that I had known and comprehended the site very well. The site was at my fingertips!

场地特质

THE CHARACTERISTICS OF THE SITE

中国园——流芳园这块场地位于汉庭顿的中西部一个山谷中央的低洼处。雨季时，山谷中会形成一条溪流，由北流向南，经过这块洼地并形成一个池塘，之后继续向南经过一个落差流过现有的日本园直到汉庭顿的西南角。低洼处面积约 10 亩，可改造成为水体，以湖沼聚全园之气，构成中国园的中心区。它的地势高度比位于南边的日本园中心水面高出约 12 米，两者相距约 100 米。低洼处的东边紧邻园内一条道路，西边是一个山丘，高约 8 米。位于低洼处北边有一块平坦的场地，面积约 3 亩，再往北是一片起伏的松林和一条溪流，地势向北逐渐高起，整个高差约 6 米。这一片区的地势是北高南低，低洼处被东、西、北三面的丘林环绕，南面视线较为开阔并由一个小溪谷向下连接到日本园。整个环境形成了一个较为理想的“风水”宝地，也基本符合《园冶》书中所说的“山林地”，并具备“有

The site for the Chinese Garden, Liu Fang Yuan, is located on the western side of the Huntington property. It is in a lower spot in the middle of a valley which runs in a north-and-south direction. Whenever it rains, a ditch in the valley becomes a stream which runs through this spot and makes it a pond. The stream continues to run through the Japanese Garden to the southwest corner of the Huntington property. The area of the lower spot is about one and a half acres large. If this area becomes a pond or lake, it would be suitable to become the center of the Chinese Garden because water, either a pond or a lake, is an essential element in a typical Chinese garden. The elevation of this area is about twelve meters higher than the pond in the Japanese Garden down the south. The distance between these two areas is about one hundred meters. A driveway is on the eastern side of this lower area, a hill eight meters high on the western side, a flat area with a half acre on the northern side, and a gentle slope towards the further northern area with a height difference of six meters. The topography of this area is higher on the north and lower on the south. The lower area is surrounded with hills and woods, and looks like a basin. Only the south side of this basin is an open flat area with a gully running down the south towards the Japanese Garden. The overall characteristics of this site are like so called “a site among wooded hills” according to Ji Cheng’s definition in *Yuan Ye*. The area has “high and hollow terrains, winding and deep spaces, tall overhanging cliffs and flat level ground, all kinds of trees reach up to the sky, flowers cover the ground. The site has its own natural attractions without the touch of human handiwork”. According

流芳园场地中心区
The central area of Liu Fang Yuan site

高有凹，有曲有深，有峻而悬，有平而坦，有成天然之趣，不烦人事之工”（《园冶·山林地》）的特点。按照计成的观点，“园地惟山林地最胜”（《园冶·山林地》）。所以汉庭顿这块“山林地”可以称得上是一处建造中国园林的绝佳场地！

南加州的气候和生态环境与中国不大相同，特别是与江南的四季分明、气候温润的环境很不一样。所以这里的植被也就不同于江南地区的植物。流芳园场地周围的植被主要有加州橡树、杉树、松树、樟树、茶树、野桃树、桉树以及部分杂树和小竹林。很少有纯中国品种的乔木。特别是围绕着低洼地，即未来中国园的中心湖区，几乎全是土生土长的加

to Ji Cheng's definition, “the most beautiful site for a garden is among wooded hills”. After a thorough investigation of the site, therefore, I had found that this site was absolutely an ideal place for the Chinese Garden — Liu Fang Yuan.

The climate and ecological environment in Southern California is quite different from China, especially from the Jiangnan (Suzhou) region where the four seasons are distinguished and the climate is humid. The vegetation or plants at the Huntington are different from those in China as well. Most large plants surrounding the site of Liu Fang Yuan were California oak trees, American

州橡树。在项目开始时，有部分人士认为这些非中国品种的树木可能不适合存在于中国园中，应该被移植或砍掉。但经过仔细勘察后，我发现这些橡树姿态俊秀、树形优美、苍劲古朴，俨然有中国文人所追求的古拙萧瑟之美，与中国园林乃至中国山水画中的树木形态基本一致，所以我决定保留部分橡树并将它们融入整个园林的景致中。

在场地的东边，有一组高大的柏树和雪松，翠绿成荫。再向东则与汉庭顿的主入口区相连，所以这组树林若稍加梳理便可形成一个顶好的入口环境，可作为整个流芳园的东入口区。位于场地东北处，有一块斜坡，生长着一片竹林，虽然长得还不太高，但未来可形成一个竹林景区。位于低洼处（湖区）的正北边是一处由一丛古老的橡树围合而成的一个约 3 亩大小的平地，每棵橡树都有各自的姿态和韵味，所以基本上可以把它们保留下来，并且可以围绕它们构筑庭院。在这块平地以北是一大片坡地，上面生长着高大的松树林，在微风的吹拂下，真有“听涛”的意境。场地的西边是一个较陡峭的坡地，其上生长有几株高大的松树和两三株美国梧桐树。这些树木都可以被保留下来并能融入流芳园中。《园冶·相地》云："旧园妙于翻造，自然古木繁花”。流芳园的场地当中能有这些姿态各异的大树，为造园提供了优越的自然环境，真是一件幸事！

在这块场地当中，最具特色的地形地貌在南边。从低洼地向南，原来是一个道路交叉口，

cedars, sycamores, pine trees, eucalyptus, wide cherry trees as well as a few Chinese camphor trees on the southern end of the site. There were also some shrubs such as bamboo and camellias on the site. It was hardly, otherwise, to find Chinese trees on the site. Particularly, in the basin area that is to become the central area or lake area of the garden in the future, there were all indigenous California oak trees indeed. At the beginning of the project, some people had suggested cutting or removing those oak trees because they were not Chinese trees. But after a closer look at them, I found that those oak trees were vigorous and beautiful in terms of their un-controlled shapes and outstretched boughs, which I thought they were as picturesque as the ones in Chinese landscape paintings and gardens. Those big oak trees presented the beauty of quaintness and desolateness which traditional Chinese scholars were seeking. I also thought that those big trees would be very good for the garden because they were old and native to the site. So in the end, I decided to save most of those oak trees as well as the other big trees on the site for the garden, and to integrate them into the design of the garden.

On the eastern side of the site, there are a dozen large cypress and cedar trees. This area connects to the main entrance of the Huntington. So to make it the eastern entrance to Liu Fang Yuan without putting a lot of efforts should be a proper decision. There is a grove of bamboo next to this area on the north. Although the bamboos were small, eventually they will grow up to a

其南边有一座木亭，木亭的一半架在石头垒成的“峭壁”上，高差大约七八米。再向南是一个溪谷，溪涧长约百余米，直通日本园。溪谷两旁主要是香樟和橡树，灌木主要是山茶树。这里有曲有深，有峻有峭，天然之趣犹存。溪谷南端有几株樱花树和梨花树散植在香樟林中，春天来临时，鲜花盛开，别有景致。

经过对现场及周边环境的详细勘察之后，我越发感到汉庭顿流芳园的这块场地具有得天独厚的自然环境，以及天造地设的造园优势和特征。事实上，在“相地”的过程中，我

bamboo forest which may become a scenic area in the garden. On the northern side of the site, there is a forest of pine trees. One can hear the “sound of waves” of the pine branches in a wind. On the western side of the site, there are a few tall pine and sycamore trees that free stand up on the hill. All these trees are saved and preserved inside the Chinese garden site. Ji Cheng said in *Yuan Ye*: “An old garden will be superb after renovating, and the old trees and various flowers are certainly to be preserved and taken advantage of.” Those wonderful trees on the site of Liu Fang Yuan will make a superior environment for the garden. This is truly a blessing to the garden!

流芳园场地北边橡树

A large oak tree in nouthern side of Liu Fang Yuan site

就已经开始对整个流芳园的构思立意进行思考和推敲了。首先是确定了流芳园的整个用地大小为 12 英亩（约 75 亩），其次是划定了用地范围，随后便开始了总体布局工作。吾师陈从周先生曾提出“园以景胜，景以园异”。那么如何才能将汉庭顿的中国园“构”成一个“景胜”且“景异”的传统式中国园林呢？我的设计工作便紧紧围绕着这块场地来“做文章”，灵活地运用中国造园美学理论和法则，把“巧于因借，精在体宜”、“景到随机”、“诗情画意”和“宛自天开”作为设计工作的最高目标。

The most distinguished feature of the site of Liu Fang Yuan is on the southern side. There was an intersection of drive ways on the western part of the Huntington, where a wood pavilion sits on a rockery “cliff” about eight meters deep. And a stream runs down to the south through a gully towards the Japanese garden. There are oak and camphor trees as well as camellias along both sides of the gully. This area is steep, winding and deep. There are a few cherry and plum trees in the gully. All these look natural and unique, especially in spring time when the flowers are blooming and the stream is cascading down.

After all these surveying and reconnoitering of the site, I had felt that the site for the Chinese garden is richly endowed by nature and a perfect place for building the garden. It embodies the characters and advantages for making it a Chinese garden. In fact, during the “situating” of the garden on the site, I had started to work on the design of Liu Fang Yuan at the same time. At first, I had defined the boundary and the area of the garden to be twelve acres. And then, I had started the master planning. My mentor Chen Congzhou once said: “A garden will be superior with its sceneries and views; the sceneries and views should be distinct in different gardens” So, how to make Liu Fang Yuan a traditional Chinese style garden with superior sceneries and views? How to make the sceneries and views of the garden different from other Chinese gardens? My design work was focused on these issues, and at the same time, on applying the Chinese aesthetics and garden principles to the design of Liu Fang Yuan. The ultimate goal was to design the garden with the concepts of “following and borrowing from the existing scenery and lie of the land, and being suitable for the site with right scales to each other”, and that “though man-made, it should look like something naturally created”, and of the “poetic and picturesque charm” as well.

流芳园场地东边橡树林
A oak grove in the eastern area of Liu Fang Yuan site

流芳园场地北侧松树
A large pine in the nouthern area of Liu Fang Yuan site

玉带接秀阁，萧疏橡树影

Jade-ribbon bridge connects the delicate pavilions, sparse oak trees scatter their shadows on the ground.

流芳园 – 橡树林与石桥、楼阁融为一体

LYF - the integration of oak trees, stone bridge and the pavilion

老蓮洪綬

第四篇

造园美学

Chapter Four

The Aesthetics of Chinese Garden

流芳园 – 题景砖雕
LFY - brick carved name board

中国园林是中国艺术中的一种独特形式，是一个将中国人的生活方式与建筑、文学、书画、戏曲、山水和花木融为一体的生命世界，是缩天地于一园的创造。中国的艺术创作，无论是诗文、绘画还是造园，首先讲究的是“意在笔先”。造园之立意，必然要涉及中国文化中的哲学思想与美学原理，必然要传承中国园林的园林艺术与造园法则，必然要反映造园者的文化情趣与精神追求。

中国园林从有文献记载开始，已经有三千余年的历史。最早记载造园活动的是《诗经》，记载了周文王的灵台、灵囿、灵沼等园林。之后，文献中记载的代表性园林有春秋时期吴越的梧桐园和会景园，秦汉时期的上林苑和甘泉苑，两晋南北朝时期的金谷园和兰亭苑，隋唐时期的西苑和神都苑，宋元时期的艮岳和琼华岛，以及明清时期的北京皇家园林“三

The Chinese garden is a unique art form in the world of Chinese arts. It combines the arts of architecture, landscape and gardening, literature, calligraphy and painting, traditional opera as well as botany and planting to create a vital world and to assemble them into a Chinese way of living. In the creation of a Chinese art, either a poem or a garden, the first step is to conceive before actually doing it. To have the conception for a garden, the designer would inevitably be involved in the realms of Chinese philosophy and aesthetic principles, and inherit the traditional Chinese garden art and its principles. It is necessary that the garden reflects and represents the cultural taste and spiritual pursuit of the master and designer of the garden.

According to the earliest recorded Chinese gardens in the historical documents, the Chinese garden has a history of about three thousand years. The earliest Chinese garden features were recorded in *Shi Jing* or the *Book of Poems* written in Chun Qiu or the Spring and Autumn Period (770-476 BC), in terms of Spiritual Platform, Spiritual Garden and Spiritual Lake in Emperor Zhou's gardens. The later Chinese documents recorded that there were many famous gardens throughout the history of China, including the Wu Tong Yuan or the Garden of Chinese Parasol Tree and Hui Lu Yuan in Wu Yue period (494—306 BC); the Shang Lin Yuan and Gan Quan Yuan in Qin and Han period (BC 221—AD220); the Jin Gu Yuan or Golden Valley Garden and Lan Ting Yuan or Orchid Pavilion Garden in Jin period (AD 265—420); the Xi Yuan or West Garden and Shen Du Yuan in the Sui and Tang period (AD 581—907); the Gen Yue Garden and Qiong Hua Dao and Tai Ye Chi

苏州留园
The Lingering Garden in Suzhou

山五园”（香山、万寿山、玉泉山、畅春园、静宜园、静明园、颐和园和圆明园）和以苏州园林为代表的江南私家文人山水园林（如沧浪亭和拙政园等）。所有这些有记载的和现存实例的园林，都直接地反映了中国园林艺术三千余年的发展变化和历史传承，它们诠释了各个时期中国人的人生哲学、美学情趣和造园艺术，以及中国艺术的核心命题：生命精神和心灵超越。

中国园林文化和造园艺术与中国哲学和美学思想紧密相连，内容十分丰富。这里选择一些与园林文化和造园艺术直接相关的中国哲学美学概念以及造园思想作一简要的介绍，

gardens in the Song and Yuan period (AD 960—1368); and the Beijing imperial gardens of "three mountains and five gardens" (Xiang Shan, Wan Shou Shan, Yu Quan Shan, Chang Chun Yuan, Jing Yi Yuan, Jing Ming Yuan, Yi He Yuan and Yuan Ming Yuan), as well as the private scholar gardens in the Jiangnan region such as Cang Lang Ting and Zhuo Zheng Yuan in Suzhou in the Ming and Qing period (AD 1368—1911). All these gardens, recorded or existing ones, have directly presented the evolution, continuity and history of Chinese garden within the last three thousand years. They have annotated the Chinese philosophies, aesthetic tastes and the art of garden making as well as the essential subject of the Chinese arts — the spirit of life and the transcendence of soul.

以便更好地理解中国园林艺术的精神所在，园林形态背后的美学思想，园林景观之外的文化寓意。

The culture and art of Chinese garden is inseparable from the Chinese philosophies and aesthetics, which have permeated the Chinese culture and civilizations in the past five thousand years. The substances of the Chinese philosophies and aesthetics are rich and profound. Here in this book, only a few subjects or concepts which directly relate to the culture and art of Chinese garden are selected and presented in order to help the readers and visitors of the garden to understand the essence of the art of Chinese garden, the aesthetic conceptions behind the outer appearances of the garden as well as the cultural metaphors embodied in the scenery of the garden.

流芳园 － 花窗、秋色
LFY - Lattice window and maple foliage

苏州拙政园

The Humble Administrator's Garden in Suzhou

生命精神

THE SPIRIT OF LIFE

《周易》一书，被誉为“群经之首”和“大道之源”，是中国最古老的哲学智慧结晶之作。《周易》所阐述的基本思想就是世界的变化和生生不息的现象，即天地万物的生命精神。中国古人相信“万物有灵”。《易传》曰：“天地之大德曰生”；“生生之谓易”；“一阴一阳之谓道；继之者善也，成之者性也。”又曰：“乾道变化，各正性命。”由周易思想而发展出来的老子思想同样视生生变化为万物的本源。《老子》曰：“道生一，一生二，二生三，三生万物”，“天下万物生于有，有生于无”。中国的艺术与哲学视生命为宇宙的根本精神。

现今著名的中国哲学美学思想家和园林鉴赏家朱良志先生在其《中国艺术的生命精神》一书中说：“中国哲学可以说是一种生命哲学，以生命为宇宙间的最高真实”，“生命是

The book of *Zhou Yi* or the *Book of Change* is seen by the Chinese as the top one in the Classics and the origin of Daoism. It is the oldest treatise on Chinese philosophies, written in Western Zhou dynasty (1046—771 BC). The fundamental ideas in the *Book of Change* are that the universe is always changing and endless, and that all things have the spirit of life. The ancient Chinese believed that all things have their spiritual lives. The *Book of Change* said: “The virtue of the universe is generating” and “going round and round is called the change”. It also said that “*Yin* and *yang* is called the Dao; the inherited one is the kind, and the one making it is the natural”. The Daoism generated from the concepts in the *Book of Change* also believed that the essence of all things is changing. Lao Zi (571—471 BC) said: “The Dao gave birth to the One; the One gave birth successively to two things, three things, up to ten thousand”, and “all things in the universe are the results of Being, and Being itself is the result of Not-being”. In the world of the Chinese art and philosophy, life is considered as the essence and spirit of the universe.

Zhu Liangzhi, a preeminent scholar in Chinese philosophy and aesthetics and connoisseur of Chinese gardens in China today, said in his book of *The Spirit of Life in Chinese Arts*: “The Chinese philosophy can be seen as the philosophy of life. Life is the very truth of the universe.” “Life is a spirit that permeates in the world of human relations, and life is the quality of creation.” Zhu also believes that “to the Chinese, life is the nature of all things including art. Art is the art for human

一种贯彻天地人伦的精神，一种创造的品质”，“在中国人看来，生为万物之性，生也为艺术之性。艺术是人的艺术，表现的是人对宇宙的认识、感觉和体验，所以表现生命是中国艺术理论的最高准则”，“生命被视为一切艺术魅力的最终之源”。

中国哲学的功用在于提高精神的境界。中国人的生命观，是天地大自然中的一切都有生命和活力。中国美学寻求的是生命的感悟和心灵的安顿。中国艺术以传达生命精神为目的，追求的重点不在形式美上，而是在生命的寓意上。中国的生命哲学思想最集中地体现在两个概念上，即“天”与“人”。天，涉及天地万物、自然、生命、时空、道、气等。人，涉及心灵、仁义、德行、性情、精神和生命体验。这些思想在《周易》、《老子》和《庄子》、《论语》和《中庸》等中国古代经典中都有所阐述，其中的生命精神蕴含在“生生不息”、“时空合一”、“天人合一”和“无往不复”等基本概念当中，它们直接影响了中国艺术

being, and the art presents the understanding, feeling and experience of human being about the universe. Therefore, to present the essence of life is the highest principle in the theories of Chinese arts. Life can be seen as the ultimate source for all the artistic charms".

The effect of Chinese philosophy is to enhance one's spiritual realm. The Chinese concept of life is that all things in the universe have their own lives and vitalities. The Chinese aesthetics is to present the comprehension of life and the comfort of soul. The ultimate goal for the Chinese arts is to convey the spirit of life so that the emphasis is not on the art form but the implication of life. There are two key concepts in the Chinese philosophy: *tian* or Heaven and *ren* or human being. *Tian* involves all things in the universe: nature, life, time and space, Dao and Qi, etc. *Ren* involves soul, benevolence and righteousness, virtue, temperament, spirit and the experience of life as well. All these thoughts and concepts are elaborated in those Classics such as the *Book of*

水仙图（局部） 宋 赵孟坚 天津市艺术博物馆藏

Zhao Mengjian (1199-1267), Narcissus(Partial), Tianjin Art Museum

包括园林艺术的内容与形式。中国园林是一种综合性空间艺术，它运用叠石、理水、建筑、花木、陈设、雕刻、诗文、绘画和戏曲等多种艺术手段和元素，构建出一个意蕴深邃、内涵丰富的生活环境。中国园林以其真实而有限的空间环境体现出造园者的生活意趣和人文理想，表达出主人的哲学思想、艺术情趣和人生追求。中国园林是中国哲学和美学思想的完美体现。

流芳园 – "得月"对联局部
LFY - "Having the Moon"

生生不息

中国人的"生生不息"概念具有多重含义。首先，"生"的本义有三个：一为孳生成长；二为生命本性；三为生之意象。概括地讲，就是天地万物的生命本性。其次，"生生不息"是天地万物之间在时空上的关联和变易。《周易》曰："是故《易》有太极，是生两仪，两仪生四象，四象生八卦，八卦定吉凶，吉凶成大业。是故法象莫大乎天地，变通莫大乎四时。"再者，"生生不息"在中国文化和哲学思想中具有一种独特的人伦含义，即儒家生命哲学的"生之为仁"概念。北宋思想家周敦颐说："天以阳生万物，以阴成万物。生，仁也；成，义也。"南宋理学家朱熹认为："仁是天地之生气"，"万物皆备于我"。所以，人必以仁心诚意去体悟生生世界，才能"合内外、同天地"，即达到儒家所强调的"上

Change, *Lao Zi* or *Dao De Jing* or *Classic of the Way and Virture*, *Zhuang Zi*, the *Analects*, and the *Moderation*. Among those thoughts, the key concepts are the "Endlessness", the "Unity of Time and Space", the "Unity of Heaven and Man", and "Circling". Those concepts have affected the contents and formation of Chinese arts including the art of garden-making. The art of Chinese garden is a synthetic artistic form which combines almost all kinds of art forms and elements including the construction of rockery and water course, architecture, botany, interior decoration, carving and sculpturing, literature and painting as well as traditional opera, etc. Combining all these elements together, a living environment with rich and profound implications is created. The Chinese garden with its actual living spaces reflects the sentiments, ideals, delights and dreams of the owner and designer of the garden. The Chinese garden, therefore, is an ultimate reflection and embodiment of the Chinese philosophy and aesthetics.

The Endless

The concept of "endlessness" to the Chinese has various meanings and implications such as birth and growth, life, and imagery of life. Endlessness is the essential nature of the universe. It reflects the connections and changes among the things in the universe. The *Book of Change* said:

下与天地同流”的最高境界，从而达到“仁”的道德高度。天地之理，生生不息而已。

中国园林是一个生命的世界！一花一世界，一水一性情，一石一精神。南宋真德秀《南康曹氏观莳园记》曰：“天壤间一卉一木，无非造化生生之妙，而吾之寓目于此，所以养吾胸中之仁，使盎然常有生意，非以玩华阅芳为事也。”朱良志先生说：“天地以‘生’为精神。”园林亦以“生”为精神。中国园林中的植物，一是重姿态，讲究自然中有生机，苍古中有生意；二是重时令，如：春柳，夏荷，秋桂，冬梅；三是重象征寓意，梅兰竹菊之“四君子”，松竹梅之“岁寒三友”。植物的选用体现了“生之态”、“生之意”和“生之趣”。园林中的“山”和“水”同样表现了生生不息的气象：山石之态彰显出恒久之精神、形影之变化；曲水之形映照出天地之清影、景色之灵动。园林建筑亦有生机，亭台楼阁，

"Therefore in the *Change*, there is the supreme ultimate at first. The supreme ultimate gives birth to the two primary forms *yin* and *yang*, the primary form give birth to the four basic images, and the four basic images to the eight trigrams. The eight trigrams can determine good fortune or disaster. Determination of good fortune and disaster can help people to accomplish great causes. Therefore, in simulation of natural phenomena, nothing is greater than the Heaven and the Earth; as for change and transformation, nothing is more obvious than four seasons." Furthermore, in traditional Chinese culture and philosophy, endlessness reflects the human nature of kindness. According to the Confucian philosophy, to love is that one can be unified with nature and be in harmony with the Heaven. The reason in the universe, therefore, contains the nature of endlessness.

花鸟画　清　陈洪绶　故宫博物院藏
Chen Hongshou (1598-1652), Bird and Flower painting,The Palace Museum

高低错落，虚实相生。整个园林是一派生趣盎然、气韵生动的世界。

时空合一

“时空合一”是中国人的宇宙观。《淮南子·齐俗训》曰：“往古来今谓之宙，四方上下谓之宇。”“宇”是空间，“宙”是时间。中国人把“宇”与“宙”合而为一来描述

The Chinese garden is a world full of living things. A flower is a world; a body of water has a temperament; and a rock has a spirit. Zhu Liangzhi said: “Heaven and earth embody the spirit of life.” A garden contains the spirit of life as well. Inside Chinese gardens, people see the vitality through plants, the endlessness of seasonal changes such as the willow tree getting green in spring, the lotus flower blossoming in summer, the osmanthus flower releasing its fragrance in autumn and the wintersweet blossoming in winter. Plants, to the Chinese, have metaphoric meanings as well. For example plum, orchid, bamboo and chrysanthemum represent four gentlemen; and pine,

临流独坐图　宋　范宽　台北故宫博物院藏

Fan Kuan (d. after 1023), Sitting Along and Facing the Water, National Palace Museum

流芳园 – 假山叠水

LFY - Rockery and cascade

表达出一个无限的世界，将时间和空间合为一体，“再造一种生命的秩序”（朱良志）。清代画家恽寿平论到艺术创作时说：“意象在六合之表，荣落在四时之外。”超越时空成为中国艺术家的追求。

中国绘画、书法、诗文、音乐、园林等艺术受到这种思维方式的影响，创造出了独具中国艺术魅力的艺术境界。中国山水画以其二维空间艺术表达出时空合一的“四维”艺术效果。南朝宋画家宗炳在《画山水序》中提出了著名的透视法则：“竖画三寸，当千仞之高；横墨数尺，体百里之远。”至魏晋以降，中国山水画的远取其势、近取其质的核心得以确立，而空间视觉原则也得以逻辑地深化。五代时期画家荆浩在《笔法记》中提出了中国山水画的中心全景模式，开创了“大山大水，开图千里”的画风。他的绘画视点极其丰富，真有“上突巍峰，下瞰穷台”的大空间，同时展现出主次分明、错落有致、远近高下、虚实藏露相结合，勾绘出一幅宏大的天地气象。北宋画家董源以皴为美的山水画，进一步在一个二维的平面中表现出三维感。从这一意义上说，笔墨即为美，即为空间。北宋画家郭熙在《山水训》中指出：“山有三远，自山下而仰山巅，谓之高远；自山前而窥山后，谓之深远；自近山而望远山，谓之平远。”此“三远构图法”使中国山水画的时空理念日趋完善。中国山水画中的“时空合一”不仅表现在绘画的构图和笔墨当中，而且展示在观赏的历程当中。画

bamboo and plum represent the three friends of winter. Similarly, the hillock and waters in Chinese garden also present the endlessness of the long lasting nature of rockery and the reflection of the sky and light in the water. The buildings in the Chinese garden present the vitality of change as well as the changes of scattered forms and appearances of light and shadow. The Chinese garden is a place full of joyfulness and verve.

The Unity of Time and Space

The notion of the "unity of time and space" is intrinsic in Chinese cosmology. It describes the unlimited universe and "re-creates an order of life", said Zhu Liangzhi. To the Chinese artists, transcending the time and space is the ultimate goal for their art works. Influenced by this philosophy, the Chinese arts including painting, calligraphy, literature, music and garden have their uniqueness and attractions. For example, the Chinese painting has a "four dimensional" effect within a "two dimensional" art form through the techniques of changing scale of the objects, lifting the view point and using different brush strokes for textures. The way to look at a Chinese landscape painting also requires moving the view point up and down or right and left, or unfolding a scroll painting from right to left. This process makes the experience of appreciating the painting in a timely manner.

面视角由多点透视法构成，所以观赏时需要平视和平移，这一过程犹如一次走进画里的“旅程”，将空间与时间结合为一体去体验山水世界中的真意和人情。观赏山水画长卷更会有此感受。

“时空合一”突显了四时之外的恒久，山川无尽，天地永恒。唐代诗人张若虚的《春江花月夜》道出了这时空的永恒：“江畔何年初见月，江月何年初照人。人生代代无穷已，江月年年望相似。不知江月待何人，但见长江送流水。”李白的《把酒问月》同样传达了时空的合一：“今人不见古时月，今月曾经照古人。古人今人若流水，共看明月应如此。”中国园林更是一个真实的“宇宙”，一个“时空合一”的生命体，一幅立体的山水画。清人戴熙云：“群山郁苍，群木荟蔚，空亭翼然，吐纳云气。”朱良志先生说：“一座小园、一座空亭，却要揽尽四海风云，宇宙灵气，让生命之流从空亭微园中流淌。园林乃空间之艺术，却要展现时间流动的韵味，时空合一，以时统空，方能极尽园林之妙。”在中国园林里，强调“虽由人作，宛自天开”的境界。这种境界蕴含了生命的空间，天地、山水、花木和建筑构成的人与景物交融的空间，也蕴含了延绵的时间，昼夜时辰、四季天象和人与物的更迭。在这样一个时空合一的园林世界里，涵泳天地的生生之韵和生命精神。

Throughout the history, in the Chinese literature, especially in the poetry, the notion of unity of time and space is always a subject to which poets express their sentiments of the everlasting. Zhang Ruoxu (AD 660—720), a poet in Tang dynasty, wrote in *A Moonlit Night on the Spring River*: “Who by the riverside first saw the moon arise? When does the moon first see a man by the riverside? Ah, generations have come and passed away; From year to year the moon looks alike, old and new. We do not know tonight for whom she sheds her ray, but see the river sending its water adieu.” Li Bai (AD 701—762), a Tang poet, wrote in *Holding Drink to Ask the Moon*: “Men in our time do not see the ancient moon, but this moon has shined on the men of yore. Men pass away like water, now as before, and all see the moon that remains for evermore.” The Chinese garden is, in fact, a real cosmos with life forms which unify the time and space, and is like a three-dimensional landscape painting. Zhu Liangzhi said: “A small garden, a pavilion, contains the spirit of the universe and the vitality of life. Garden design is a spatial art that presents the charm of changing time and unifies time and space. Through time, a garden space can become a wonderful place.” The Chinese garden, though a man-made environment, appears as something naturally created. It comes into a realm of life, an integration of the sky, the earth, the plant and the building, a place where the changes of time, day and night, seasons and people occur. Inside a Chinese garden, therefore, the endlessness of universe and the vitality of life have all been presented here.

秋山晚翠图　五代　佚名　台北故宫博物院藏
Anonymous (ca. 910-960),
Autumn Mounation and Evening Green,
National Palace Museu

天人合一

“天人合一”是中国文化和中国哲学的基本精神，它体现了中国人对待宇宙与人生之间关系的核心观念。最早明确提出“天人合一”观念的是汉代大思想家董仲舒，他在《春秋繁露》中明确提出：“天人之际，合而为一”，并且认为“仁之美者在于天”，“举天地之道，而美于和”。北宋思想家张载《正蒙·乾称》曰：“儒者则因明致诚，因诚致明，故天人合一，致学而可以成圣，得天而未始遗人。”意思是说，人与天之间因明诚而融为一体，最终达到天人合一的境界。天人合一的思想早在西周时期已初见端倪，到春秋战国时期儒家和道家都追求天人合一的理想，只是两家所采用的途径不一样。儒家用“仁”、“性”和“诚”的概念诠释了“天”与“人”合而为一的思想。《易传》提出“与天地合其德”，《中庸》讲“尽人之性”，“尽物之性”，《孟子》则讲“尽性知天”，“上下与天地同流”。朱熹亦曰：“惟仁然后能与天地万物一体。”其基本思想是通过提高人的道德心性修养，达到与天合一的境界。道家以“道”和“自然”的概念说明“人”与“天”合而为一的思想。老子《道德经》曰：“人法地，地法天，天法道，道法自然。”庄子《齐物论》曰：“天地与我并生，而万物与我为一”，“人与天一也”。

The Unity of Heaven and Human Being

The essential spirit of Chinese culture and philosophy has embodied in the notion of the unity of Heaven and human being. Dong Zhongshu (179—104 BC), a prominent thinker at Han dynasty, was the first to define the notion in a more meaningful way. In *Spring and Autumn Studies*, Dong said: “The - Heaven and man shall be made one”, “the kindness is the beauty of Heaven”. Under the influence of Confucius, Zhu Xi (AD 1130—1200), a philosopher in Song dynasty, said: “Only love and kindness can make man unified with the Heaven and the Earth.” The Confucian emphasized the quality of man's mind and heart. The Daoism, on the other hand, expresses the notion of the unity of Heaven and man through nature and the Dao. Lao Zi said in *Dao De Jing*: “The ways of man are conditioned by those of earth; the ways of earth by those of Heaven; the ways of Heaven by those of Dao, and the ways of Dao by those of nature.” Zhuang Zi said in *The Equality of Things*: “Heaven and earth coexist with me, and all things are unified with me as a whole”, and “man and Heaven are in one”.

中国古代哲学家惠施也说："泛爱万物，天地一体也。" 惠施还提出了"至大无外，谓之大一"的概念，即每个人、每个事物都应当被看作宇宙的一部分。董仲舒进一步阐述"天人合一"思想的内涵，他提出："取天地之美以养其身"（《循天之道》），"天地人，万物之本也。天生之，地养之，人成之，不可一无也"（《立元神》）。关于"天人合一"的思想，中国人普遍的一种理解是"人"与"自然界"的和谐统一。而另一种理解则认为"人心就是天"，"天内在于人"，正如宋代哲学家陆九渊所云："我心即宇宙，宇宙即我心"。

Hui Shi (370—310 BC), an ancient Chinese philosopher, said: "Love all things, and then Heaven and earth are unified as one". It meant that all things including human being are integral parts of the universe, so we should love all things in the universe. Dong Zhongshu said: "Heaven, earth and human are the origin of all things. Heaven created everything, earth nurtured it and human being made it meaningful. All three are inseparable from one another." The meaning of "unity of Heaven and human being" as comprehended by the Chinese has two aspects, one is that nature and human being is in harmony with each other; second, the Heaven is in a person's heart, as Lu Jiuyuan (AD 1139—1193), a Song dynasty philosopher, said: "My heart is the universe, and

梧竹秀石图　元　倪瓒　北京故宫博物院藏

Ni Zan (1301-1374), Bamboo and Rock, Palace Museum

流芳园 – 太湖峰石

LFY - Taihu Peak Rock

竹石图　元　管道昇
台北故宫博物院藏
Guan Daosheng (1262-1319), Bamboo and Rock, National Palce Museum

天人合一是人的心灵与自然的对话。"当一个人感到已经没有你我之分、物我之别，感到宇宙就是自身、自身就是宇宙时，他的精神就达到了最高的满足。这种境界就是天人合一"（引自姜义华《中华文化读本》）。中国禅宗的"不二法门"哲学，强调"平等不二"和"无分别"，从而达到了"一"的最高境界。由此可以看出，"天人合一"思想包含着物质世界与精神世界两个方面，即"天"与"人"在物质世界中和谐，在精神世界里合一。

中国园林艺术同样追求"天人合一"的境界。中国当代哲学家李泽厚和刘纲纪在《中国美学史》中指出："'天人合一'、'天人感应'、'天人相通'，实际上是中国历代艺术家所遵循的一个根本原则。"事实上，中国园林就是一个"人"与"天地"合一的生活居所。中国园林提供了一个理想的有限空间，让人与宇宙大自然生生不息、四时运转地融为一体，达到物我同得素心、合而为一的境界。这种"合一"的境界反映出人的情感与自然界的相应关系，正如西晋文学家陆机在《文赋》中所云："悲落叶于劲秋，喜柔条于芳春"，以及北宋哲学家程颢在《秋日偶成》诗中所描述的那样："万物静观皆自得，四时佳兴与人同。"园林是"为情而造景"，"情景交融"即是人与物的交融，即所谓"我见青山多妩媚，料青山见我应如是"（辛弃疾《贺新郎》）。人与天和，即人与自然的和谐，人与自然不争不取，一任自然。陶渊明的"此中有真意，欲辨已忘言"，亦是一种物我合

the universe is my heart." It emphasizes the communication between human spirit and nature or universe. Therefore, the notion of "unity of Heaven and human being" has both the physical as well as the spiritual implications, namely physically heaven and man is in harmony with each other, while spiritually they are unified into one.

The art of Chinese garden seeks the unity of Heaven and man as its ultimate goal. Li Zehou and Liu Gangji said in their book of the *History of Chinese Aesthetics*: "Pursuing the unity of Heaven and man, the interaction and communication between nature and human being is the essential principle to the Chinese artists throughout the history." A Chinese garden, in reality, is a living place where the Heaven and earth is unified with man. With a pure spirit and heart, the designer of a Chinese garden creates an ideal garden within a limited space to achieve the integration of nature, time and man. The unity of Heaven and human being presents the direct relationship between man's sentiment and the change of nature. Lu Ji (AD 261—303), a scholar of Jin dynasty, wrote in Wen Fu: "A sadness falls as the leaves fall in the late autumn; the happiness raises as the soft branches grow in the spring." Cheng Hao (AD 1032—1085), a poet in Song dynasty, wrote in a poem: "I quietly look at the things and find they are all contented with themselves; man thrives along with the four seasons." A garden is created for the sentiment of man, nature and man as well as the things and man have been in harmony with each other. "There are

一的境界。中国造园强调“宛自天开”，也就是强调“人作”与“天造”合而为一的境界。

无往不复

中国哲学中的生命精神还表现在另一个概念上，即 “无往不复”。无往不复的核心思想是“循环无限”和“四时运转”，这是一种中国人特有的时间观。《易经》曰：“无平不陂，无往不复”，《易·象传》曰：“无往不复，天地际也。”《彖传》曰：“‘反复其道，七日来复’，天行也。复：其见天地之心乎！”可以看出，《易传》把“无往不复”作为宇宙生命的根本表现形态，在时间与空间的结合与交替，构成了宇宙“变易”的特性。《易·系辞下传》曰：“日往则月来，月往则日来，日月相推而明生矣；寒往则暑来，暑往则寒来，寒暑相推而岁成焉；往者屈也，来者信也，屈信相感而利生矣。”《老子》曰：“有物混成，先天地生。寂兮寥兮，独立而不改，周行而不殆，可以为天地母。”自然界中，物由生到衰，再由衰到生；日夜不停地运转，四时永恒地更替，都是循环往复的。中国人以无往不复的时间观去体验和把握自然规律，从而延展自己的精神空间和开拓自己的心灵世界，通过诗文、绘画和园林等艺术的表达，来满足这些陶冶情操的精神欲求。中国园林

essential meanings in nature, but it is beyond my words to describe them", wrote by Tao Yuanming (AD 365—427), a poet of Jin dynasty. This situation described in the poem presents the real unity of heaven and human being. The Chinese garden emphasizes the effect of being naturally created, which means unifying the man-made and the natural elements in the garden into one.

Circling

Another concept in the traditional Chinese philosophy is the notion of "circling" or circulation of things in the universe. The notion of circling is a Chinese concept for time. It is said in the *Book of Change* that "no one goes forward without return". Lao Zi said: "There was something formless yet complete that existed before heaven and earth; Without sound, without substance, dependent on nothing, unchanging, all pervading, unfailing. One may think of it as the mother of all things under the Heaven." In the universe, all things are from birth to decline, yet from decline to re-birth, and go on and on. Day and night re-circling, the four seasons replace each other endlessly. All these present the notion of circling in nature. The Chinese people see and understand the universe and human nature from this perspective. They extend their spiritual and mental spaces into a higher level through various art forms including poetry, painting as well as garden, where one can experiences the changes of seasons, time of day and night, the circling of the sun and moon, etc. In

桃花鸳鸯图　宋　佚名　南京博物院藏
Anonymous (Song d.), Peach Flower and Mandarin Ducks, Nanjing Museum

流芳园 – 秋荷
LFY - Autumn Lotus

为此提供了一个理想的空间让人体察无限生命往复的形式。这种往来循环和有限与无限统一的观念，在中国园林艺术中也有所体现，春天看柳，夏日观莲，秋天赏桂，冬日寻梅。老树开花、枯木逢春；观四季轮回，察日月运转，“一一得其时宜”。

每当置身于中国园林之中，在优游寄趣之时，在畅神达性之刻，中国哲学和艺术中的“生命精神”便能够得到充分的体现。空间上，园林灵在“曲折”、“通透”、“幽深”和“流动”中；景象上，园林妙在“虚实”、“光影”、“动静”和“声色”上；景物上，园林活在“曲水”、“顽石”、“佳木”和“云烟”间；寓意上，园林借在“诗情”、“画意”、“比德”和“寄托”里。中国园林就是一个生生的世界！

short, everything has its own timing.

When strolling in a Chinese garden, enjoying the scenes and satisfying the sentiment, one may understand the traditional Chinese philosophy of the spirit of life. Inside a Chinese garden, the spaces are winding and flowing; the views are illusive, vivid and colorful; the scenes are various including bridges, rockeries, plants and pavilions in the mist....; the metaphoric and poetic meanings are embodied in the scenes and literary expressions. The Chinese garden is a world of vitality.

人文山水
THE CULTURAL LANDSCAPE

高阁观荷图　宋　佚名　朵云轩藏
Anonymous (Song d.), Looking at the lotus from the Tall Pavilion, Duo Yun Xuan

中国艺术的生命精神，最直接地体现在中国的山水文化上。中国人的山水文化已经延绵了几千年，山水情结已经融入中国人的生活当中。上古时期有对自然的崇拜；夏、商、周时期有祭祀天地的灵台、灵沼和灵囿；秦汉时期有范山模水的“蓬莱仙境”和“一池三岛”；魏晋时期有“隐逸山林”和“玄对山水”的田园山居；隋唐时期有“寄情山水”的别业草堂；宋元时期有“中隐司官”；直到明清时期有“城市山林”的文人山水园林。可以说，亲近山水、融入山水、寄情山水是中国人“天人合一”思想的直接体现和执著追求。从中国传统文化角度看，这些山水不是完全客观的天然山水，而是被提炼与人格化了的山水，是被赋予了

The spirit of life embodied in the Chinese arts has been directly represented in the cultural landscape of China. This culture has lasted for thousands of years. The deep emotions associated with the landscape have been integrated into the lives of Chinese people. In the ancient times, the Chinese worshiped nature; in Xia, Shang and Zhou periods (2070—256 BC), the Chinese built terraces, ponds and gardens to practice ritual activities for spiritual connections to the Heaven and the earth; in Qin and Han dynasties, there were the “Penglai Wonderland” and “A Immortal Lake and Three Islands” inside the imperial gardens; in Wei and Jin dynasties, there were hermit cottages in the mountains and woods and pastoral dwellings in the countryside; in Sui and Tang dynasties,

丰富人文内涵的山水，是人心灵中的山水。中国文化中的山水是人文的山水，是生存的感悟，是心性的写照，是感情的寄托，是心灵的慰藉！最著名的人格化山水比喻就是孔子的“知者乐水，仁者乐山”。先民与自然一开始便用艺术的方式对话，自然的探索与艺术的鉴赏滋润着人们的心与眼，让一个天人合一的中国图式由混沌而至清晰。中国的山水文化承载着中国人文领域里最为深邃的寄托。

自然界中的山山水水，在中国文化中被作为人之心性本源的表达方式和寄情抒怀的对象。自然山水所带来的气息是清纯的、真实的和自然的，人融入山水之中而获得心灵的净化和本真的回归。“文人未有不好山水，盖山水远俗之物也”（清代李果《墨庄记》）。“寄身于自然之中，与山水相融一体，是‘齐物’的通达途径，亦即纵游山水乃是为进入‘齐物’或‘从心所欲不逾矩’的境界”（邵琦《中国画文脉》）。与此同时，这些山水因为人的

对月图　南宋　马远　中国台北故宫博物院藏
Ma Yuan (1140-1225), Looking at the Moon, National Palace Museum

there were villas residing in the cultivated landscapes; in Song and Yuan dynasties, there were reclusive mansions in the cities; and in Ming and Qing dynasties, there were scholar gardens of the “urban forest”. All these have shown that the interest in getting closer to the landscape, being integrated with the landscape and keen on the landscape is the direct reflection of the Chinese belief of the “unity of Heaven and human being”. From the perspective of the traditional Chinese culture, the *shan shui* or mountains and waters or the landscape is not exactly the same as the nature appears to be; instead, it is the cultivated and personalized landscape. The landscape in Chinese culture is the humanized cultural landscape, an understanding of nature, a portrait of mind’s eye, an expression of the sentiment, and a comfort of soul! One of the most famous personalized expressions of the landscape is Confucius’ saying in the *Analects*: “The wise delights in waters, the benevolent delights in mountains.” The ancient Chinese communicated with nature through arts. Along with the exploration of the natural world as well as the artistic creation, the Chinese has made the imagery of “the unity of Heaven and human being” more realizable and clear. The cultural landscape has been the most essential sustenance in the humanities of the Chinese culture.

The natural landscape has been the object to the Chinese to express their thoughts of mind and sentiments of heart. Li Guo (AD 1679—1751), a Qing dynasty scholar, said in *Mo Zhuang Record*: “There is no scholar who is not fond of the mountains and waters, for the reason that the mountain and water is something away from the vulgar things.” Shao Qi, a contemporary artist, said in his book of the *Context of Chinese Paintings*: “Being in nature and integrated with nature is the

参与而具有人文的气质和内涵。在中国历史上，寄情山水、养闲林泉、归居田园的生活方式是文人士大夫的高尚追求。东汉隐士严子陵避居于富春江畔，东晋书圣王羲之修禊于会稽山兰亭，唐代诗人王维闲居于辋川别业，北宋诗人林逋隐居于西子湖畔的孤山。这些文人雅士登山临水，隐逸林泉之中，一是可以脱离世俗环境，从而可以悟道和净化心灵；二是借自然山水以抒发林泉之致，同时又赋予山水以人文气息，使这些山水具有了历史意义和文化内涵。比如，当人们面对富春江，即想到严子陵，想到高风亮节，想到垂钓；进入会稽山，即想到王羲之，想到曲水流觞，想到《兰亭集序》；游览西湖孤山景区，便想到林逋，想到泛舟，想到仙鹤。此时的山山水水已经不再是纯“天然”的了，而是人文的山

way to be equal with all things. And touring in the natural landscape, one will reach the state of being equal with all things and follow one's inclinations." Throughout the Chinese history, scholars always pursue the life style of being in natural landscape, cultivating in the landscape and living in pastoral countryside. Yan Ziling, a hermit of Eastern Han dynasty (AD 25—220), lived a secluded life on the bank of Fu Chun River; Wang Xizhi (AD 303—379), the prominent calligrapher of Eastern Jin dynasty, cultivated in Lan Ting in Kuai Ji Mountain; Wang Wei, a poet of Tang dynasty, resided in Wang Chuan Villa in Lantian; and Lin Bu (AD 967—1028), a poet of Northern Song dynasty, secluded in Gu Shan on the bank of the West Lake in Hangzhou. These scholars not only

兰亭修禊图卷（局部） 明 仇英 故宫博物院藏

Qiu Ying (1498-1552), Lan Ting Gathering (partial),The Palace Museum

垂纶图 清 陈洪绶 故宫博物院藏
Chen Hongshou (1598~1652), Fishing, The Palace Museum

水！北宋大画家郭熙把山水比喻为“春山淡冶如笑，夏山苍翠如滴，秋山明净如妆，冬山惨淡如睡”，正像是辛弃疾所写的“我看青山多妩媚，料青山见我应如是”，这是一种人与山水的合一与对话。明代张洪《耕学斋图记》曰：“人不得山水，无以畅其机；山水不得人，无以显其奇。”可以看出，中国文化是如此强调人与自然的融合，人与天地的合一。人与自然之间充满着生命气息和万物之情。

中国的山水文化主要包括山水诗文、山水绘画和山水园林。中国艺术讲“道法自然”，讲“外师造化”，讲“宛自天开”。山水诗派的鼻祖南朝宋谢灵运的诗充满着道法自然的精神，比如在《登池上楼》中写春天：“池塘生春草，园柳变鸣禽”；在《初曲郡》中写秋色：“旷野沙岸净，天高秋月明”；在《岁暮》中写冬景：“明月照积雪，朔风劲且哀”。唐代诗人王维和孟浩然的山水诗创作更达到了前所未有的高峰。“声喧乱石中，色静深松里”（王维《青溪》诗句）；“山光忽西落，池月渐东上。荷风送香气，竹露滴清响”（孟浩然《夏日南亭怀辛大》）。借山水以栖身心，山水已成为参悟人生意义和宇宙真理的手段。山水情怀在元代散曲当中得到了悠扬的咏唱，成为山水文学中的一朵奇葩。关汉卿《白鹤子》：“鸟

secluded themselves from the society and purified their minds and souls through the landscape, but also left their marks on the landscape with aesthetic and cultural meanings. Guo Xi (AD 1000—1080), a preeminent painter and theorist of Northern Song dynasty, said: “The Spring mountains are so light and elegant as a smile, summer mountains so greenish as if they were dripping water, autumn mountains so clear as with a makeup, and winter mountains so gloomy as if they were sleeping”. Xin Qiji (AD 1140—1207), a poet of Southern Song dynasty, said: “I can see how charming the green mountains are, at the same time, I think the mountains are looking at me in the same way.” This presents the unity and dialogue between the landscape and human being. Zhang Hong of Ming dynasty said in *Geng Xue Studio Record*: “One can’t express freely ones ideas without being with the landscape; the extraordinariness of the landscape would not be noticed without the involvement of human beings.” In Chinese culture, therefore, the focus is on the integration and unity of human being with nature. Between human being and nature, there are full of the spirit of life and the love of all things.

啼花影里，人立粉墙头。春意两相牵，秋水双波溜。香焚金鸭鼎，闲傍小红楼。月在柳梢头，人约黄昏后。”白朴《天净沙·春》：“春山暖日和风，阑干楼阁帘栊。杨柳秋千院中，啼莺舞燕，小桥流水飞红。”卢挚《秋景》：“挂绝壁松枯倒倚，落残霞孤鹜齐飞。四围不尽山，一望无穷水。散西风满天秋意。夜静云帆月影低，载我在潇湘画里。”这些山水诗词曲，从描写自然山水景色到人与景的融合，从“无我”之境到“有我”之境，再到“情景交融”，人文山水尽显其中。

中国的“山水画”几乎就是中国画的代表，中国画中有近百分之七十为山水画。中国画又有“文人画”之称。北宋文豪苏东坡最先提出“文人画”或“士人画”的概念，并强调以“取其意气”为重，同时倡导“诗情画意”的文人画风格。中国山水画也是中国“文人画”的代表，它不带有宗教和政治色彩，而全然表达出人的心性和对自然天地的真实感悟。中国山水画中的“山水”也不是完全模仿自然的山水，而是画家将眼所观之山水与心所悟之山水相结合，通过笔法自然地流露于纸上。唐代画家张璪《绘境》曰：“外师造化，中得心源。”他所指出的是绘画创作要以自然天地为师，取之于自然，同时又要经过画家本人心灵的感悟和陶冶，方能表现出作品的艺术生命力即神韵。现存最早的山水画是隋代展子虔的《游春图》。南朝宋画家宗炳的《画山水序》被称为中国第一篇山水画论。宗炳曰：

The Chinese cultural landscape or the *shan shui* culture has been manifested in many artistic forms such as in landscape poetry, landscape painting and landscape gardens. The Chinese art places its core value on the importance of learning from nature, following the natural course as well as making the art work seem to be naturally created. Xie Lingyun (AD 385—433), a poet of Southern dynasty and the first landscape poet in China, wrote his poems with the spirit of naturalism, for example in *Deng Chi Shang Lou* describing the spring scenery: “Spring grass grows in a pond; the willows in the garden whisper like songbirds”; in *Chu Qu Jun* describing the autumn scenery: “Sands on a wild river bank are clean; the autumn moon high in the sky is bright”; and in *Sui Mu* describing the winter scenery: “The bright moon illuminates the snow; the north wind is strong and doleful.” Wang Wei and Meng Haoran (AD 689—740) were two preeminent landscape poets of Tang dynasty. Their poems reached a higher level of landscape poetry. “Rapids hum over heaped rocks, while colors are set silently in the thick pines”, are lines from *Qing Xi* by Wang Wei. “The sunset suddenly goes down on the west from the mountain top; the moon rises above the pond from the east… Wind with lotus fragrance sends out aroma, drips of the dews from bamboo make clear sounds”, are lines from a poem of Meng Haoran. The temperament and interest in landscape is also expressed in the verses of Yuan Qu or the Opera of Yuan dynasty. Guan Hanqing (AD 1234—?), a preeminent opera writer of Yuan dynasty, wrote beautiful verses in the *White Crane*: “Birds sing songs in the shadow of flowers; people stand by the white washed walls. The awakening of spring

“山水以形媚道，而仁者乐”，“夫以应目会心为理者，类之成巧，则目亦同应，心亦俱会。应会感神，神超理得。”与宗炳同代的王微，亦是纵情丘壑的人物。他在《叙画》中提出：“望秋云，神飞扬；临春风，思浩荡”。南朝著名画家谢赫在《古画品录》中提出的中国绘画艺术标准的“六法”中的第一条就是“气韵生动”。“气韵生动”是品画的最高美学准则，需要表现出所画对象的生命精神和内在特征，即对象的“神采”和“风韵”。五代时期画家荆浩《山水节要》曰：“意在笔先，远则取其势、近则取其质”。被后世画界称之为“马一角”的宋代画家马远，善于以小见大。他不用常规画法的全景式构图，而是采用一种主

ties in with those two; the double waves slip through the autumn water. Incenses are burning in a duck-shaped tripod, while people are relaxing near the red house. The moon sets on the treetops, while people make dates after the evening.” All these lines and verses described the sceneries as well as the integration of human being with the natural landscapes. These descriptions include the “no me”, and “with me” sceneries and the blending of the sentiments with the sceneries. The landscapes are clearly the cultural landscapes.

The Chinese landscape painting almost stands for the Chinese painting. The landscape paintings account for about 70% of all Chinese paintings. Another name for the Chinese painting

杂画四（局部） 明 文徵明 故宫博物院藏

Wen Zhengming (1470-1559), Landscape (partial), The Palace Museum

观式的截景或框景方式，省略掉一切与主题无关的元素。中国画家这种高超的提炼自然的手法，妙造自然，一切都是为了弘扬主旨，一切都成就了主题的衬托。心境跃然于尺幅之间，意趣流露在点画之外，巧妙地把心目中的山水尽情地摹写出来。

中国园林的审美理念服从于中国绘画的审美情趣。“中国造园首先从属于绘画艺术”（童寯，当代著名建筑师）。中国绘画创作追求“外师造化，中得心源”，中国园林创作讲究“虽由人作，宛自天开”。中国园林中如果没有“山”，没有“水”，就不成为“园林”。然而，中国园林里面的山是“借”来的：其一是借自然之石之土人为地创造的山，称之为“假山”；其二是借园外之真山，将其“收入”园内，构成园景。中国园林中的水也不是“自然而然”的水，而是“理”出来的水：其一是引自然之水入园内；其二是人为地将水曲折变化，水随山转。但“范山模水”和“叠山理水”还需做到“宛自天开”，才能不失真山水之面目，达到“心源”的山水与“自然”的山水合而为一。

中国古代的隐逸文化对中国的山水文化和园林文化具有直接的影响。隐逸文化作为一种“出世”文化的形式，强调“返璞归真”和“高蹈遁世”的意识，以“天人合一”作为最终追求目标。隐逸文化源自老庄的哲学思想，“贤者伏处大山嵁岩下”（《庄子·在宥》），“山林与，皋壤与，使我欣欣然而乐与”（《庄子·知北游》）。东晋诗人陶渊明《饮酒》曰：

is called “scholar painting”, first named by the prominent scholar Su Shi (AD 1037—1101) of Northern Song dynasty. The scholar painting focused on the presentation of the artistic intention and the poetic meaning. The landscape painting, in most cases, does not represent any religious or political implications. They present the disposition of the artist and comprehension of the natural world. The landscape in a Chinese landscape painting is not an imitation of natural landscape, but a combination of the images in the eye and mind of the artist. Zhang Zao (AD ?—1093), a painter of Tang dynasty, said in *Hui Jing* or the *Realm of Paintings*: “Learn from nature, but create from the heart of the source.” The first theoretic book on Chinese painting was the *Preface of Landscape Paintings*, written by Zong Bing (AD 375—443), a painter of Southern dynasty. He said: “The landscape reflects the Dao or the Way through its charming form and texture, so the benevolent delights in the landscape. We comprehend the *Li* or truth based on what we see in our eyes and how we feel in our hearts. If a work of art is articulated well, the creator and viewer of the work will see and feel the same thing. When this happens, the quintessence of *Li* or truth will be comprehended.” Jing Hao (AD 850—?), a painter of Five Dynasties, said in his book of the *Summary of Landscapes*: “One should conceive before starting to paint and capture the posture of faraway objects and the texture of the nearby things.” Ma Yuan (AD 1140—1225), a painter of Song dynasty, usually drew not to the panoramic views of the landscape, but only a few elements, and put them on a corner of the painting. This unique painting style was called the “Ma’s Corner” by the Chinese. The Chinese

山水册　明　陈洪绶　故宫博物院藏
Chen Hongshou (1598-1652),
Landscape, The Palace Museum

"采菊东篱下，悠然见南山。"唐代诗人白居易在《中隐》中提出了"'小隐'隐山林，'中隐'隐司官，'大隐'隐市朝"的概念。隐逸山水文化从强调地理空间环境向着心灵空间环境发展，从而使得"山水"从自然环境走向人居环境，乃至进入室内环境。于是山水画、山水楹联、山水盆景、顽石奇石便进入庭院和文房之中，成为中国人特别是文人雅士生活中不可分离的内容，即所谓"会心处不必在远"。"晋人向外发现了自然，向内发现了自己的深情"(宗白华《美学散步》)。在与自然顾盼之间，艺术家拟人化了山石湖泊，艺术生命由此与山川大地融为一体。这种"中隐"文化思想造就了中国园林的"城市山林"特征。"城市山林"

painters created these theories and techniques in order to capture and represent the essence of the landscape to express their thoughts and sentiments in the paintings, and beyond.

The aesthetic conceptions and principles of the Chinese garden are similar to those in the Chinese painting. "Learning from nature, creating from the heart as the source" is the doctrine of the Chinese painting, while the doctrine of the Chinese garden is that "though man-made, it should look like something naturally created". To the Chinese, if there is no "mountain" and "water", it should not be called a garden. Inside a Chinese garden, however, the mountain is "borrowed"

甚至成为中国园林的代名词。现存最早的“城市山林”是宋代诗人苏舜钦建于苏州的沧浪亭，它开启了苏州园林以“中隐”为内涵的“城市山林”特征，而苏州拙政园又是“城市山林”的一个代表性杰作。“明四家”之一的文徵明所绘的《拙政园图咏》中的 31 幅画册就生动地描绘了“城市山林”的景象。在文人的诗句中对“城市山林”也有所描述，如沧浪亭的“一径抱幽山，居然城市间”（苏舜钦《沧浪亭》），拙政园的“绝怜人境无车马，信有山林在市城”（文徵明《拙政园图咏》），狮子林的“谁谓今日非昔日，端知城市有山林”，“疑

because, first, it is a man-made hillock or mountain which is called "*jia shan* or artificial mountain"; second, it may be located outside the garden, but "borrowed" into the garden scenery. The pond or lake inside a Chinese garden is not naturally formed but manipulated by the human being through for example, drawing water from outside into the garden or making the existing water more winding and circling. While imitating the natural mountain and water, the landscape in the garden should be made like they are naturally created. It should not be lacking of the essence of natural landscapes. It should be able to unify the landscape conceived in the mind's eye and the one from nature.

The seclusion culture in ancient China has directly influenced the culture of landscape and

晴峦萧寺图（局部） 宋 李成 美国纳尔逊艺术博物馆藏

Li Cheng (919-967), The Clear Mountain and Temple, The Nelson Art Museum

苏州拙政园

The Humble Administrator's Garden in Suzhou

苏州狮子林
Lions Grove Garden in Suzhou

其藏幽谷，而宛居闹市。假山似真山，仙凡异尺咫”（清乾隆《游狮子林》）。清代顾汧《凤池园集》曰：“若夫结庐人境，心远地偏，不出郊埛，而神已游濠濮之表，盖境不自异，因心而开，非所称城市山林者耶？”这种“城市山林”的艺术空间，“真正实现了中国文人历来所渴慕的‘结庐在人境，而无车马喧’的最高美学理想”（金学智《中国园林美学》）。和美宁静的自然山水，是古人最终的心灵归宿，可谓“人入造化之中，尽得造化之趣”。王羲之《兰亭诗》曰：“三春启群品，寄畅在所因。仰望碧天际，俯磐绿水滨。寥朗无厓观，寓目理自陈。大矣造化功，万殊莫不均。群籁虽参差，适我无非新。”

gardens. To renounce the world, the secluded seeks to return to innocence and to be in a hermit kingdom, so as to reach the highest state of the unity of Heaven and human being. The origins of the seclusion culture came from the philosophic thoughts of Lao Zi and Zhuang Zi. Zhuang Zi said: “Mountains and forests, hills and plains, all of them make me rejoice.” Tao Yuanming of Jin dynasty wrote lines: “While picking chrysanthemums by the eastern fence, my gaze upon the Southern Mountains rests.” Bai Juyi (AD 772—846) of Tang dynasty once said about the seclusion: “The one who secludes in mountains and forests is called the little hermit; the one in a city is called the moderate hermit, and the one in imperial court is called the super hermit.” The focus of the seclusion culture had shifted from being in natural landscape to urban environment, so that one can enjoy the convenience of daily life in the city while retreating to the mountain, water and forest in

中国山水文化及其山水园林是中国人生命精神的体现，是情思、心性和心灵的寄托，也是提高心性修养和悟究天地之道的途径，更是天人合一思想的写照。中国园林以“真山真水”的形式将人文内涵诠释其中。宗炳在《画山水序》中提出：“山水质有而趣灵”、“山水以形媚道”。郭熙在《林泉高致》中提出的山水画有：可行、可望、可游和可居的特征，而中国园林的山水是实现这“四可”的真实空间，不仅如此，园林更是满足精神生活的场所，是实现孔子所追求的“游于艺”的理想天地。

the garden. The culture of seclusion had become a part of the lives of the scholars. The Chinese said: “An intimate place does not need to be far.” The so called “moderate seclusion” in urban environment had made the Chinese garden a *cheng shi shan lin* or “urban mountain and forest”, a name sometimes is substituted for the Chinese garden. This phenomenon had become popular in the Jiangnan region, especially in Suzhou area. The oldest and existing example of the “urban mountain and forest” garden in China is Cang Lang Ting or the Garden of Surging Wave Pavilion, first built in Song dynasty in Suzhou. Most gardens in Suzhou including the Humble Administrator’s Garden and the Lion Grove Garden have the characteristics of the “urban mountain and forest”. This type of garden has made it possible to realize the ideal way of life, “where people reside, I build my cot, the noises of horses and wagons I heard not”, verse by Tao Yuanming. To the ancient Chinese, a harmonious and serene landscape is the ultimate destination of the mind and soul.

流芳园 － 双桥连翠

LFY - Two bridges connet the Greens

沧豁图　明　文徵明　故宫博物院藏
Wen Zhengming (1470-1559), The Ravine , The Palace Museum

The cultural landscape and scholar gardens in China have been the showcases of the Chinese spirit of life. They are the places where people can express emotions, dispositions and thoughts of their minds. They are the ways through which people can improve themselves and their comprehensions of the truth of nature. They are also the depictions of the ideology of the unity of Heaven and human being. The Chinese garden annotates the connotation of Chinese humanities through its formation of the real mountains and waters. Zong Bing said in the *Preface of Landscape Paintings*: "Mountains and waters have the quality of inspiration. The landscape reflects the Dao or the Way through its charming forms and textures." Guo Xi said: "The landscape painting has the characteristics that one can walk into it, look at it, stroll in it and live within it." The Chinese garden, in fact, is a real space where one accomplishes those four activities. The Chinese garden is, in addition, a place where one can evoke the spiritual life. It is also an ideal place to realize the pursuit of what the Confucius wanted, "to play within arts".

造园法则

THE PRINCIPLES OF GARDEN-MAKING

中国园林通过真实的四维世界将中国人的生命精神和审美情趣表现出来，壶公天地、芥子世界，表现了大化生机的世界。中国园林艺术创造的最高原则是“虽由人作，宛自天开”（计成）。中国的造园艺术，从早期的“借山借水”，到“范山模水”，再到“宛自天开”，经历了三千余年的历史，演变和发展出了中国的造园思想和基本法则。它们集中地体现在以下六个方面：因地制宜、小中见大、有无相生、巧于因借、诗情画意、宛自天开。这些造园法则相互交融，互为补充，从而使中国园林成为丰富多彩、气韵生动、清秀典雅、寓意深厚的人居环境和精神家园。

The Chinese garden is a four-dimensional cosmos full of vitality and delight. The most important principle of Chinese garden construction is as Ji Cheng said: "Though man-made, it should look like something naturally created." The Chinese garden has been evolved from "borrowing from nature" to "imitating nature", then to "naturally created" in the last three thousand years. The cardinal principles of Chinese garden design and construction have been developed and advanced throughout the history. The following six concepts are the epitome of them: fitting in with the site, seeing the large from the small, growing out of being and not-being, following and borrowing from skillfully, being poetic and picturesque, and looking like naturally created. These principles should complement each other in order to create a Chinese garden where one can experience a rich, vivid, elegant and meaningful living environment and be in a spiritual home.

Fitting in with the Site

The primary principle of garden design and construction is to make the garden "fit in with the site". There are three aspects: the first is to situate the garden on the site; second, to follow the form and contours of the site; and third, to make the garden suitable for the site. Ji Cheng said in the chapter of *xiang di* or situating in *Yuan Ye*, "The foundation of a garden should not be restricted to any particular direction; the land will have its natural highs and lows. The gate to the garden should be attractive; the scenery should follow the natural lie of the land, whether the garden lies beside

因地制宜

造园的首要原则是“因地制宜”。因者，依据；制者，制定；宜者，合适。在造园当中，其内涵包括三个方面，一是相地，因地；二是随形，随势；三是合宜，得体。《园冶·相地》云：“园基不拘方向，地势自有高低；涉门成趣，得景随形，或傍山林，欲通河沼。”其中的关键概念有以下几方面：一是“因”：利用所处地域特点，布置园景。如“高方欲就亭台，低凹可开池沼”。二是“随”：景到随机、得景随形、随曲合方、随势因借、随宜合用。三是“宜”：精而合宜、精在体宜。合宜，合适之意，与环境相宜才合用。

wooded hills or abuts on a river or pond." The key at the beginning of a garden design, therefore, is to find out the characters and special features of the site and to make best use of them and befit each other. For example, it is proper to put a pavilion and terrace on the higher ground, and to dig a pond on the lower spot. And then, scenes should be created in accordance with the spatial characters of the site. After that, appropriate scales and proportions should be applied to the structures.

苏州拙政园

The Humble Administrator's Garden in Suzhou

《园冶·兴造论》云："因者，随基势之高下，体形之端正，碍木删桠，泉流石注，互相借资；宜亭斯亭，宜榭斯榭，不妨偏径，顿置婉转，斯谓'精而合宜'者也。"所以，造园首先是要选择场地，一旦选定合适的造园场地之后，则需详细"相地"，从而"因地制宜"地布置园林景观。"因地制宜"就是要真实地反映和利用场地的地形地貌特征，无论山林地、城市地、村庄地、郊野地、傍宅地还是江湖地都各有其特质，将园林各个部分有机地结合到场地当中，做到随宜合用，精而合宜，景到随机，让园林自然地从场地"生长"出来，呈现出天配地适、宛自天开的意境效果。

始建于北宋，苏州沧浪亭是现存最古老的苏州园林，以"因地制宜"手法造园而著称。苏州宅园大都为封闭式，园墙高筑，内外隔绝。沧浪亭则依山傍河而建，园外借山，园前借水；复廊环山而绕，挑临水面；"沧浪亭"坐落山巅，四周古木参天，山石嶙峋；厅堂水榭，高低错落。整个园林古朴素雅，有若自然，与苏州小桥、流水、人家融为一体，相得益彰。

In the "Theory of Construction" in *Yuan Ye*, Ji Cheng said: "Following the lie of the site: laying out the garden in accordance with the rise and fall of the natural contours as well as the spatial forms, lopping trees or branches that block the spaces for structures, using rocks to direct the flow of a spring, so that each benefits the others. Wherever it is suitable for a pavilion, build a pavilion, and wherever a gazebo, build a gazebo. Paths should be hidden away and winding with the land; this is the so called 'delicacy and appropriateness'." Selecting a suitable site for a garden, therefore, is the first task to design and construct a garden. In this regard, a meticulous survey of the site should be done beforehand in order to find out the characteristics of the site. And then, the garden ought to fit in with the site as the garden is "growing" from the site. In this way, the garden will look like something naturally created from the site.

First built in Northern Song dynasty, the Garden of Cang Lang Ting or the Surging Wave Pavilion is the oldest existing garden in Suzhou. It is well known for its fit with the site and surroundings. The gardens in Suzhou are usually connected with the residences which are enclosed with tall walls and separated from outside. This garden was built beside a canal and situated next to a hill. The water of the canal and the wooded hill became an integral part of the garden sceneries. In addition, the pavilions, gazebos and halls as well as rockeries inside the garden were laid out in accordance with the site, so the garden becomes very unique and attractive. It is integrated with the urban context of Suzhou very well.

苏州留园
The Lingering Garden in Suzhou

小中见大

宋僧道灿曰:“天地一东篱,万古一重九。”元代张宣在题倪云林《溪亭山色图》中曰:“江山无限景,都聚一亭中。”“小中见大”,是中国艺术特别是园林艺术创作的一条重要原则。无论园林建造得多大,范围都是有限的。中国人有“天地为庐”的宇宙观。“纳千顷之汪洋,收四时之烂漫”;“一瓢日月,十笏山河”。在这有限的空间里,采用“移天缩地”的手法,创造出一个“壶公天地、芥子世界”。壶虽小,天地却很宽,因为这“壶”是心灵之壶。中国园林就是一个“壶公天地”,在有限的空间里,构造出一个大乾坤,一个生命的世界。

Seeing the Large from the Small

Zhang Xuan (AD 1341—1373), a poet of Yuan dynasty, wrote an annotation on a Ni Zan's painting: "Unlimited views of a landscape, all gathered into one pavilion." "Seeing the large from the small" is an important principle in Chinese arts, especially in the art of Chinese garden. No matter how large a garden is, it would be constructed within a limited space. The Chinese cosmology believes that the essence of the universe or heaven and earth can be comprehended in a hovel. A garden contains a view of a watery expanse of thousand hectares and the changing

唐代李白《下途归石门旧居》诗："何当脱屣谢时去，壶中别有日月天。"杜甫《春日江村》诗："乾坤万里眼，时序百年心。""

"小中见大"，就是心性的延展，从而诠释人与自然、人与宇宙的关系。老子曰："不出户，知天下。不窥牖，见天道。"中国文学和艺术常常运用"以少胜多，小中见大"的艺术手法，诗词的句子以简练来表达无尽意境，中国艺术所表现的空间意识可概括为"以小见大"和"以大观小"。宋代苏辙《洞山文长老语录》："古之达人，推而通之，大而天地山河，细而秋毫微尘，此心无所不在，无所不见。是以小中见大，大中见小，一为千万，千万为一，

brilliance of the four seasons. The sun and moon can be seen in the water inside a dipper; the landscape can be represented inside a ten-yard-wide garden. The art of Chinese garden is to create, within a very limited space, a so-called gourd of heaven and earth, and a vivid world of life.

"Seeing the large from the small" is an effect of the extension of mind's eye. It announces the relationship between human being and nature as well as the universe. Lao Zi said: "Without leaving his door, he knows everything under heaven; without looking out of his window, he knows all the ways of the heaven." In Chinese literature and art, the principles of "less is more" and "seeing the large from the small" are often applied to make abundant artistic conceptions and spatial effects. Su Shi, a Song dynasty poet, said: "The wise man in ancient times knew many things, from the

苏州狮子林

The Lions Grove Garden in Suzhou

流芳园 － 月门洞

LFY - A Moon Gate

皆心法尔。”中国绘画史上也有“小中见大”的画家，如南宋的马远和夏圭。马远的画“树取一枝，石取一角，溪出一湾”，景少而意多，物小而韵长。朱良志先生说：“对小的重视，反映了人们注重平和、悠远、淡雅的心理需求。”元代画家倪瓒的山水画最具“小中见大”、“以少胜多”的意境。元代张宣题倪瓒画诗曰：“石滑岩前雨，泉香树杪风。江山无限景，都聚一亭中。”倪瓒的画有一种无言之美：疏林挺秀，淡水沼沼，远山平卧，小亭独立。清代画家郑板桥说：“雅室何须大，花香不在多。”清代画家恽寿平《南田画跋》曰：“意贵乎远，不静不远也；境贵乎深，不曲不深也。一勺水亦有曲处，一片石亦有深处。”这些都反映出中国艺术遵循“以小见大”的创作原则。

中国园林艺术借鉴中国绘画理论，创作出“小中见大、以小观大”的园林世界。中国园林强调“一花一世界，一草一天堂”；“一叶落，知劲秋；一月圆，知宇宙”，“一勺一水以梦千寻海浪，一石一峰以梦万仞高山”。江南园林以小巧别致著称，以淡雅清逸取胜。清代沈复在《浮生六记》中讲到园林时曰：“若夫园亭楼阁，套室回廊，叠石成山，栽花取势，又在大中见小，小中见大，虚中有实，实中有虚，或藏或露，或浅或深。”在园林布局和造景中，欲小中见大，必曲折幽深，方寸得宜，楚楚有致。运用“分隔”、“曲折”和“幽深”的空间处理手法，加上“虚实”和“露藏”的造景手段，以有限面积造无限空间，创造出“小

great landscape to the micro cosmos, where his heart felt everything and his eyes saw everything. The reason was that he can see the large from the small and see the small from the large. One embodied thousands of things, while thousands of things came to one. All these were true because they came from his heart.” In Chinese paintings, for example, Ma Yuan and Xia Gui, two Song dynasty painters, applied this principle to draw only a couple of branches, a corner of a rockery, and a small bend of a stream on their paintings in order to create the much larger and meaningful and imaginable landscapes. Zhu Liangzhi said: “Paying greater attention to the small means that people look for peaceful, profound and elegant things to meet their psychological needs.” One of the best so-called the minimalist Chinese landscape painters is the Yuan dynasty recluse-artist Ni Zan (AD 1301—1374). His paintings presented a wordless beauty by composing only a grove of elegant trees, misty water, distant mountains and a solitary pavilion. Zheng Banqiao (AD 1693—1765), a Qing dynasty painter, said: “Elegant rooms need not largeness, the fragrance of flowers need not plenty.”

Similarly, the principle of “seeing the large from the small” is applied to the Chinese garden art as well. Inside a Chinese garden, a puddle of water could evoke an imagination of a thousand waves; a peak rock could represent a high mountain. The gardens in Jiangnan region especially in Suzhou, are delicate and elegant. Shen Fu (AD 1763—1825), a Qing dynasty scholar, said in

中见大”、“园中有园”、“景外有景”的效果。

比如，苏州网师园就是“小中见大”、“以少胜多”的园林典范。区区八亩余地，以一泓小池造就了池水汪洋弥漫之感；以布局紧凑创造了“园中有园”、“景外有景”的空间序列；以曲廊、小桥、石径串联庭院形成了步移景异的变化；以古柏怪石尽得了苍古之意；以青竹、碧水、秋月和寒松隐喻了四季景象；厅堂、楼阁、轩榭、馆室、斋屋、桥廊，应有尽有。优游其间，实有无穷意趣。

有无相生

《周易》里阐述的阴阳关系是最重要的对立统一论之一，用“阴”与“阳”两仪的互相依存，互相变化，揭示一个丰富多彩、变莫测的世界。《周易·系辞》曰：“一阴一阳之谓道”。老子曰：“天下万物生于有，有生于无”，“有无相生”，“有之以为利，无之以为用”。清人笪重光在《画筌》中论山水画曰：“山之厚处即深处，水之静时即动时。林间阴影，无处营心；山外清光，何以着笔。空本难图，实景清而空景现；神无可绘，真境逼而神境生。位置相戾，有画处多属赘疣，虚实相生，无画处皆成妙境。”中国园林的造园之道亦在其中。

his book of the *Six Records of a Life Adrift*: “If I build a garden, I will make pavilions and halls, winding corridors, rockery mountains, postured flowers, and create scenes that enable me to see the large from the small and to blend the real with the virtual, the hidden with the exposed, and the light with the dark.” The layout of spaces and composition of scenery in the Chinese garden are all in the matter of these skills to create the effects of seeing the large from the small, a garden within a garden, and sceneries beyond sceneries.

The Master of Nets Garden in Suzhou is a good example of application of the principle of “seeing the large from the small”. Within a limited area of 6,000 square meters, the garden was composed of a central pond which looked like a large watery and a series of courtyards and corridors creating effects of “a garden within a garden” and “views beyond the sceneries”. There are various plants including bamboos, pines and old cypresses, etc., and the changing brilliance of the four seasons is represented inside the garden as well. In addition, there are many garden structures including halls, pavilion, waterside studios, bridges and corridors within the garden. Strolling in the garden, one will experience the richness and charm presented in most Chinese gardens.

中国园林当中充满着这种对立统一、相生相济的造园手法和景观意境。陈从周先生在其园林著作中，较为完整地阐述了造园景观中的这种辩证关系。比如："动"与"静"：静寓动中，动由静出；以静观动，以动观静，则景出。"实"与"虚"：虚中有实，实中有虚，虚实相生。"露"与"藏"：山重水复，柳暗花明。"有限"与"无限"：于有限空间，造无限景致。"有形"与"无形"：以无形之诗词意境，构有形之水石亭台。"情"与"景"：以情悟物，以物抒情；情景交融，为情造景。"造"与"借"：造者，园以景胜，景因园异；借者，园外有园在于借，景外有景在于时。陈从周先生论园林时又说："'空灵'二字，为造园之要谛"，"池水无色，而色最丰"，"刚以柔出，柔以刚现"，"文贵乎气，气有阳刚阴柔之分，行文如是，造园又何独不然？"

Growing out of Bing and Not-being

In *Zhou Yi*, "yin" and "yang" elaborated the meanings of the unity of opposites. It said: "Alternation between *yin* and *yang* is called the Way." Lao Zi said: "All things under heaven are the products of Being, and Being itself is the product of Not-being." "Things grow out of Being and Not-being." "Being is for the benefit, Not-being for the usefulness." Da Chongguang (AD 1623—1692), a Qing dynasty painter, said in *Hua Quan*: "The depth of a mountain lies in its big volume; the dynamic of water lies in its tranquility. When the real images are clear, the invisible comes to life. There are wonderful and mysterious things in the empty spaces on the painting." These are also true in the art of Chinese garden.

流芳园 － 石岸、乐鱼

LFY - Rockery bank of the Lake and happy fish

在中国园林里，有无相生的景致无处不在。比如，“动”者有叠泉流水、枝叶飘拂、花影移墙；“动观”有步移景异、俯仰换景。“静”者有亭台楼阁、假山奇石、曲径梁桥；“静观”有亭中观月、峰峦当窗。“实”者有山水、屋宇、花木；“虚”者有清风、疏影、漏窗。“露”者有亭榭、粉墙、峰石；“藏”者有曲径、深潭、暗香，等等。这些说明中国园林充满着辩证思想，“有”与“无”相生相济，千变万化。

巧借于因

计成在《园冶》开篇《兴造论》中提出了“园林巧于因借”的思想，“借者，园虽别内外，得景则无拘远近。”计成在《园冶》的最后一篇《借景》中又提出：“构园无格，借景有因”，“因借无由，触情俱是”，“夫借景，林园之最要者也。如远借，邻借，仰借，俯借，应时而借。然物情所逗，目寄心期，似意在笔先”。计成阐述了借景在造园中的独特地位，以及借景的不同方式。“借景”是中国园林造园的重要手法之一。“借景”一是指借园外之景，“借”是指向不完全属自己的一方借之，即陈从周先生所说的“化他人之物为我物”，“园外有景妙在因借”。当然，园外之物并不都是好景，需“俗者屏之，嘉者收之”（计

Chen Congzhou had articulated this dialectical relationship applied in the Chinese garden with a holistic approach. For instance, regarding “in-motion” and “in-position”, he said that “in-position resides in the motion, while in-motion comes out from the in-position”, and “scenery will be created when one in a still position look at the moving features, and vice versa.” Regarding the “real” and “illusive”, he said: “the illusive is within the real, and vice versa. The two complement each other.” He said, regarding the “hidden and revealed”, “they are like “the endless mountains and waters juxtapose the shady willows and bright flowers.” Regarding “the limited” and “the unlimited”, “a garden should be built within a limited space to present the unlimited scenery.” Regarding “the formed” and “the formless”, “a garden should be built by using the formless poetry and literature to construct the formed water, rockery and pavilions.” Regarding the “sentiment” and “scenery”, “one ought to comprehend things through sentiments and to express sentiments through things; to integrate these two together, and to create scenes in accordance with the sentiments.” Regarding “the created” and “the borrowed”, “a garden will be attractive with scenes, and gardens differ from each other with its unique scenes.” Finally, Chen said: “‘*Kong ling* or openness and spiritual are the gist of garden design and construction.”

Inside a Chinese garden, there are many scenes with this kind of dialectical relationship. The “moving” scenes include the flow of spring water, floating tree branches, shadows of flowers on the wall; the “strolling-viewing”, scenes changing with different steps, or the looking-upward

苏州拙政园
The Humble Adminstrator's Garden in Suzhou

成）。夹巷借天，门窗借景，临湖借影，夜色借月。可借之物和可借之景，其实不限于通常所说的园外之景色，还包括园内之景物。借景之物可包括花草树木、亭台楼阁、湖光山色、梅兰竹菊，还有流水、远山、青天、白云、飞鸟、夕阳、明月；甚至那些无形的景象，芳香、渔歌和清影。在苏州网师园里，有一亭名“月到风来”，临池而设，得风月为园所有，这样的“因借”实属神来之笔。

"借景”的描述在中国文学作品中也不胜枚举。最早且最著名的借自然山水之景，应该是晋代陶渊明的“采菊东篱下，悠然见南山”。其他还有，“山嶂远重叠，竹树近蒙笼”（南朝沈约《游沈道士馆》），“孤帆远影碧空尽，惟见长江天际流”（唐李白《送孟浩然

and looking-downward scenes; the “still” scenes include pavilions and terraces, rockery hillocks and peak rocks, winding paths and bridges; the “in-position-viewing”, to look at the moon inside a pavilion, a landscape through a window frame; the “real” things include mountains and water, buildings, plants; the “illusive” things include breeze, shadows and lattice windows; and the “revealed” things include pavilions, white washed walls, peak rocks; the “hidden” things include winding paths, deep pond, fragrance, etc. All these present the dialectical relationship and the ever-changing essence of “being” and “not being” inside Chinese gardens.

之广陵》),"青霜红碧树，白露紫黄花"（唐王维《秋圃》），"窗含西岭千秋雪，门泊东吴万里船"（唐杜甫《绝句》），"江上晴楼翠霭间，满帘春水满窗山"（唐李群玉《汉阳太白楼》）。这些诗句反映出借景的种类是多种多样，有远借，邻借，仰借，俯借，应时借等。比如，中国园林中，"远借"最佳的实例是北京颐和园借玉泉山之妙高塔于园内，苏州拙政园借报恩寺之北塔于园中。宝塔矗立云间，塔影倒插波中，亦实亦虚，别有韵致。

Skillfulness on Following and Borrowing from

Ji Cheng said in *Yuan Ye*: "A garden should be constructed skillfully to 'follow' and 'borrow from' the existing scenery and lie of the site." "To borrow from the scenery is not limited to whether it is close by or far away though the garden has a boundary which separates the inside from the outside." In the last chapter of "Borrowing from the Scenery" in *Yuan Ye*, Ji Cheng said: "There are no fixed rules for designing gardens but there are reasons for borrowing from the natural scenery." "There is no definite way of borrowing from the scenery; it lies in whatever stirs your emotions." So "'borrowed scenery' is the most vital part in the design of a garden. The 'borrowed views' include distant view, nearby view, the view above, the view below, and views as time and seasons change. But the attraction of natural objects, both the form perceptible to the eye and the essence which touches the heart, must be fully imaged in your mind before you put brush to paper." The concept of "borrowed view" meant to borrow from the scenes outside of the garden. "To make something of somebody else become yours", and "the scenery outside of the garden should be skillfully borrowed", said Chen Congzhou. The ungainly things, of course, must be screened off. In fact, the concept of "borrowed view" is not limited to the views outside of the garden, but refers to the inside ones as well. The views that can be borrowed from include flowers, trees, pavilions, waterscape, distant mountains, the blue sky, the clouds, the flying birds, the moon, and even the formless things such as the fragrance, the rippling sound of flowing water, the songs of birds or fisherman, etc. Inside the Master's Fishnet Garden in Suzhou, there is a pavilion named as "the breeze arrives and the moon comes", which makes the breeze and moon be part of the scenery of the garden. It is truly a work of the Heaven.

In Chinese literature especially in poetry, there are many poems with the lines "borrowing from the scenery" to depict the landscape and the sentiment of the poet. Among these, the earliest and most famous verse is written by Tao Yuanming: "While picking chrysanthemums by the eastern fence, my gaze upon the Southern Mountains rests." A preeminent poet of Tang dynasty, Li Bai wrote: "A lonely sail with its distant image vanished in blue emptiness; all I see is the great river flowing into the far horizon." Du Fu (AD 712—770), another preeminent poet in Tang dynasty, wrote: "The long accumulated snow on Xi Mountains can be viewed from the window; ships from the Eastern State of Wu thousand miles away are anchored in the docks just outside the house door."

Li Qunyu (AD 813—860), a Tang dynasty poet, wrote: "A riverside tower sets in the mist of greens in a sunny day, the spring water and mountains are framed into the windows." All these presented the various ways of "borrowing from the scenery" from afar and nearby, from above and below, and from changing time and seasons. In the Chinese garden, good examples of "borrowing from the distant scenery" are the borrowed view of the Miao Gao Pagoda on the Jade Spring Mountain into the Summer Palace Garden in Beijing, and the view of North Pagoda in Bao En Temple into the Humble Administrator's Garden in Suzhou. Here, the pagodas stand high up to the cloud, their images inverted in the waves on the water. The real or illusive images make charming scenes in the garden.

苏州留园 - 湖石倒影

A reflection of a rock in the Lingering Garden in Suzhou

流芳园 - 楼阁倒影

A reflection of a pavilion in LFY

诗情画意

西汉文学家扬雄《法言》曰："言，心声也；书，心画也。"中国古代将艺术看成是人的内在思想感情的真实反映。宋代苏轼曰："诗画本一律，天工与清新。"诗歌，是中国文学中表达思想感情最简洁、清晰和丰富的一种形式；绘画，是中国艺术中传达意境最直观的一种方式。中国园林则以水木清华之美，叙述着"诗情"与"画意"。因有"诗情画意"而使中国园林更有意境。计成《园冶・园说》可以说就是一篇"园林意境篇"。计成写道：

"山楼凭远，纵目皆然；竹坞寻幽，醉心即是。轩楹高爽，窗户虚邻；纳千顷之汪洋，收四时之烂熳……紫气青霞，鹤声送来枕上；白苹红蓼，鸥盟同结矶边。"

"夜雨芭蕉，似杂鲛人之泣泪；晓风杨柳，若翻蛮女之纤腰。移风当窗，分梨为院；溶溶月色，瑟瑟风声；静扰一榻琴书，动涵半轮秋水，清气觉来几席，凡尘顿远襟怀。"

"诗情"与"画意"尽显其中！

中国园林追求"诗情画意"，包含两部分内容：其一是"诗情"，抒发思想和感情，即对生命意义的思考和人心性的流露，重在表达"境界"；其二是"画意"，表现"入画"的视觉美感和画外的"象外之象"，重在传达"意境"。概括地说，就是 "诗情"讲"境界"，

Being Poetic and Picturesque

Yang Xiong (53—18 BC), a Western Han dynasty scholar, said in *Fa Yan*: "Words are the voice of the heart, writing or calligraphy is the image of the mind." To the ancient Chinese, art was the true representation of people's sentiments. Su Shi of Song dynasty said: "The arts of poetry and painting share the same principles: being natural and creative." The Chinese poetry presents the sentiment in a very concise, clear and meaningful way, while the Chinese painting presents the artistic conception by direct visual means. The art of Chinese garden embodies both the poetic and picturesque qualities in presenting the landscape. Ji Cheng's writing in the chapter of "On Gardens" in *Yuan Ye* can be seen as a chapter of the artistic conceptions of a garden: "Gaze into the distance from a tower on the hill top, and beautiful scenes will meet your eyes. Seeking a secluded spot in the bamboo grove, one will revel in it. The corridors are tall and wide, and windows and doors give unimpeded views." "Transplant some bamboos in front of the window, and set aside some pear trees to form a courtyard; the moonlight spills in, and the wind whispers; the shadow disturbs the lute and books on the couch, and the wind ruffles the reflection of the crescent moon in the autumn water. We feel a pure atmosphere around the table and seats, and the worldly dust is suddenly far away from our minds."

苏州拙政园
The Humble Administrator's Garden in Suzhou

“画意”求“意境”。境界有高低，意境有深浅。境界与意境二者都生在象外，超越形式而具神韵，有言外之意、韵外之致、景外之景。从唐代诗画的境界说，到明清书画的意境说，都是中国传统美学中绕不开的话题。“境界”是“妙明心中之物”（南宋画家米友仁）。清代钱泳《履园丛话》中曰：“造园如作诗文，必使曲折有法。”

王国维在《人间词话》中对“境界”有着十分明确的诠释：“能写真境物，真感情者，谓之有境界；否则谓之无境界。”他强调“词以境界为最上。有境界则自成高格，自有名句。”他还提出“有造境，有写境”之分，以及“有我”之境与“无我”之境之别。中国园林同

Throughout Chinese gardens there are two aspects regarding the concept of “poetic and picturesque”. The “poetic” expresses the sentiment on the meaning of life and thoughts of mind. It focuses on the expression of the extension of the sentiment. The “picturesque” presents the visual aspect of the scenery and the meaning behind the imagery. It focuses on the conveyance of the artistic conception. Both the “poetic” and “picturesque” conveys the suggestive quality of the scenery in the garden. Qian Yong (AD 1759—1844), a scholar of Qing dynasty, said: “To design and construct a garden is the same as to write poetry. Some principles have to be applied to the story-telling.”

样以“有境界”为最上。造园也是造境，即“写”出真境物，表达出真感情。

陈从周先生在《中国园林艺术与美学》一文中指出：“中国文学、戏曲、园林都是重感情的抒发，突出一个‘情’字。” “园林是一首活的诗，一幅活的画，是一个活的艺术作品。”他又在《中国诗文与中国园林艺术》一文中进一步说：

“诗有诗境，词有词境，曲有曲境，画有画境，音乐有音乐境，而园林之高明者，运文学绘画音乐诸境，能以山水花木，池馆亭台组合出之，人临其境，有诗有画，各臻其妙。故‘虽由人作，宛自天开’，中国园林，能在世界上独树一帜者，实以诗文造园也。”

朱良志先生在《中国美学十五讲》一书中对“境界”作了精辟的阐述：

“境界，其实就是世界。是人心所对之世界，不是实在之存有，而是虚灵之世界。”

“对宇宙之觉解、生命之感悟、人生之体验，形成了人不同的境界。”

“境界标示人的意识所对之世界、人心营构之世界以及因象所观之世界，这三者贯通一体。与审美创作者相关之世界，指物境；审美创造者心灵构造之世界，指心境；鉴赏者再创造之世界，指意境。三个世界属于不同层次，具有不同的意涵，但又相通。”

金学智先生在《中国古典园林美学》中也说：

Wang Guowei (AD 1877—1927), a scholar and preeminent critic on poetry of later Qing dynasty, had articulated *jing jie* or state or realm in writing of poetry especially lyrics in his book of *Talks on Poetry in Human World*. He said: “Those poems which can describe true scenery and objects, true emotions and feelings, can be said to possess the *jing jie* (State or Realm). Otherwise they may be said to lack *jing jie*.” “The best poems are the ones with *jing jie*. If a poem has *jing jie*, it will naturally achieve a lofty style and naturally possess eminent lines.” He also said, “There is a creative *jing jie*, and there is a descriptive *jing jie*.” Similarly, the best Chinese gardens are the ones with *jing jie*. The garden design and construction is to create the scenery with *jing jie* so as to present the true scenery and objects, and true emotions and feelings.

Chen Congzhou said in his article of the *Art and Aesthetics of Chinese Garden*: “A great attention is paid to the expression of sentiment in Chinese literature, opera and garden. The emphasis is on the word of ‘sentiment’.” “The Chinese garden is a lively poem, a lively painting, and a lively form of art.” Chen also said in his article of *Chinese Poetry and the Garden Art*: “Poetry has its own state, Chinese opera has its own state, painting has its own state, and music has its own state. The skillfully designed and constructed gardens

苏州拙政园

The Humble Administrator's Garden in Suzhou

“在园林中，特别是在名园里，可以说处处蕴含着诗意，时时荡漾着诗情，事事体现着诗心，是地道的‘诗世界’。”“诗情”重“境界”。象存境中、境生象外，“以境显意”。

诗言志，亦言情。宗白华先生在《美学散步》中论及诗与画时说：“诗可以完全写景，写‘无我之境’。而每句每字都反映出自己对物的抚摩，和物的对话，表出对物的热爱，那纯粹的景就成了纯粹的情，就是诗。” 唐王维《鹿柴》：“空山不见人，但闻人语响。返景入深林，复照青苔上。”宋王安石《舟夜即事》：“山泉如有意，枕上送潺湲。”唐诗将山水风景的意趣写得美到极致，而宋词将园林的情趣写得入木三分。北宋文学家晏殊《寓意》词：“梨花院落溶溶月，柳絮池塘淡淡风。”宋代文学家欧阳修《蝶恋花》词：“庭院深深深几许，杨柳堆烟，帘幕无重数。”《朝中措·平山堂》词：“平山阑槛倚晴空，山色有无中。手种堂前垂柳，别来几度春风。”辛弃疾《满江红·暮春》词：“红粉暗随流水去，园林渐觉清阴密。庭院静，空相忆。”宋词对园林美的细腻描写与对主人情致的抒发融为一体、契合无间。园林是为人建造的，其中的景致寄寓了人的情感。诗文所写之景致、所抒之情怀，成为造园之源思。“情能生文，亦能生景”（陈从周）。中国园林不仅造景借助诗文之情境，而且用诗文题咏的方式点缀园林的景致，使园林更有“文化性”和“书卷气”。这也是中国园林艺术的特征之一，正如金学智先生所说：“中国园林极富供人栖居的诗意”。

integrate the variety of states of the art including painting and music, composing the mountain and water, the trees and flowers, the pavilions and waterside halls so that people in the scene can see and feel the poetic and picturesque scenery as well as the charm of the garden.” “Therefore, ‘though man-made, it should look like something naturally created’, the Chinese garden is so unique in the world of gardens because it is designed and constructed in accordance with the principles in poetry and literature.”

Zhu Liangzhi has annotated the notion of *jing jie* or state incisively in his book of *Fifteen Talks on Chinese Aesthetics*: “*Jing jie*, in fact, is a world where a person feels by his heart. It is not a real being but an ethereal world of one’s soul.” “Each *jing jie* is different from one another based on each person’s perception of the universe, the understanding and experience of life.” “There are three parts in *jing jie*: the awareness of the world, the structure of the world in mind, and the imagery of the world through the eyes. These three parts are unified and connected with each other. The world related to the artist is the world of physical environment; the world constructed in artist’s heart and mind is the world of mind; the world that a viewer sees and imagines is the world of conception. These three worlds are in different levels with different meanings, but they are connected with each other.”

“画意”源自中国的绘画艺术。在古代象形字起源时期“书与画”或“文与画”是同一个意思，即以形传达意思或思想感情。五代两宋是中国山水和花鸟画的兴盛期，也是中国画论的繁荣期。代表作品如五代时期关仝的《山溪待渡图》，荆浩的《匡庐图》及其山水画论《笔法记》和《画山水赋》；北宋时期范宽的《溪山行旅图》，郭熙的《早春图》及其著名的画论《林泉高致》。这些山水画所表达出的思想即“画意”，以及画论中所提出的绘画原则和手法，对自魏晋以来的文人山水园林的发展产生了巨大的影响。元明清是写意山水画和文人画的鼎盛时期，园林山水图也成为画家创作的主题之一，其中的代表作品包括元代倪瓒

Jin Xuezhi, a contemporary garden scholar, said in his book of the *Aesthetics of Classical Chinese Gardens*: “In the gardens, especially in the famous ones, there are poetic implications everywhere in the garden, poetic sentiments in all times, and poetic thoughts in all things. A garden is truly a ‘world of poetry’.” The poetic sentiment relies on the *jing jie*. Imagery resides in the *jing jie*, and *jing jie* develops beyond the imagery. *Jing jie* should be used to present the meaning.”

苏州拙政园
Tha Humble Administrator's Garden in Suzhou

流芳园 － 门洞细部
LFY - A garden gate detail

的《狮子林图》，明代沈周的《东庄图》，文徵明的《东园图》和张宏的《止园图》，以及清代宫苑画册《圆明园四十景图》等。这些不仅是山水画卷，更是园林意境的写照。

中国文化和艺术以传递人的情怀和意念为重点，即所谓“传情达意”。“画意”讲“意境”，“意境”重达意，传物象意趣。中国园林同样以传达“意境”为上。计成在《园冶》自序中云：“不佞少以绘名，性好搜奇，最喜关仝、荆浩笔意，每宗之。”造园须“宛若画意”、“想出意外”。陈从周先生在论及中国园林与山水画的关系时说：“言意境，讲韵味。表高洁之情操，求弦外之音韵，两者二而一也。”“画中寓诗情，园林参画意，诗情画意遂为中国园林之主导思想。”中国山水绘画之法讲究经营位置，注重文理；讲究写意和概括，注重“画外之音”。“画意”对造园的影响表现为两方面，一是“构图”、“入画”或“布局”，二是“写意”即“意境”。“意境”强调心灵体验、妙会，如清代沈复所说：“小景可以入画，大景可以入神。”绘画不同于诗文，绘画是运用具体的二维空间表达出画家的思想感情。中国山水画是最接近园林艺术的艺术。山水画的构图理论与造园布局原则有相通之处。宋人张淏在《艮岳记》中描写艮岳园林的壮观景色时曰：“丘壑林塘，杰若画本，凡天下之美，古今之胜在焉。”宗白华先生说：“诗和画圆满结合，就是情与景的圆满结合，也就是所谓‘艺术意境’。”正如计成所说的“顿开尘外想，拟

Poetry is an expression of mind and sentiment. Zong Baihua, a contemporary scholar on aesthetics, said in his book of *Strolling in Aesthetics*: “A poem can present scenes only, a scene without oneself in it or the no-me scene. But it actually reflects one’s feeling about the things, the communication with the things and the love of the things, so the purely scenic images become something purely sentimental. That is what poetry is about.” Wang Wei, a Tang dynasty poet, wrote lines in *Lu Chai*: “There seems to be no one in the empty mountain. And yet I think I heard his voice. Where the sunlight reached in a deep forest, it shines on the green moss.” Wang Anshi (AD 1021—1086), a Song dynasty poet, wrote lines in *Zhou Ye Ji Shi*: “If the mountain spring is willing, it will flow slowly in the dreams over the pillow.” Ouang Xiu (AD 1007—1072), a Song dynasty poet, wrote in *Die Lian Hua*: “In a deep and deep courtyard, willow trees look like a heap of mist, and curtains after curtains.” A garden is made for people, so the scenery should reflect the sentiment. The descriptions of the scenery and expressions of the sentiment in the poetry should become the source of inspirations in designing gardens. “Sentiments can inspire the writing of poetry and the making of scenery,” said Chen Congzhou. The design and construction of Chinese garden is not only inspired by the descriptions in poetry, but applied the lines of poetry to the inscribed couplets and boards on the garden features and structures. Therefore, the Chinese garden is rich with cultural elements and literary atmosphere. Jin Xuezhi said: “The Chinese garden is a poetic living place.”

入画中行”。中国园林追求的“诗情画意”或“艺术意境”，所要传达的就是人与自然之间的圆满结合，将人的情感和意念寄托在山水佳木之中，从而实现颐养性情、陶冶情操、安顿心灵和提高精神境界的目的。

宛自天开

中国造园的最高境界即是计成所提出的“虽由人作，宛自天开”。扬雄曰：“人不天不因，天不人不成。”中国艺术在天人交合的基础上谈创造。“虽由人作，宛自天开”实为造园之宗旨和最高原则。中国园林就是以人为的创造达到“巧夺天工”的境界，就是人造的“天然图画”，即计成所说的“自然天成之趣”。中国哲学思想体现出“天地与我并生，而万物与我为一”《庄子》，人与物都是宇宙的一部分，是共生与共存的关系。“人作”的园林要取法自然，顺应自然，又要概括自然，高于自然。“外师造化，中得心源”，达到人与自然合而为一，体现出“寓巧于拙，寓美于朴”的思想。

The notion of “picturesque” derived from the art of Chinese painting. During the ancient times, the Chinese characters were presented with images or hieroglyphics. So writing and painting were the same thing that were mainly used to convey meanings or sentiments. In Tang dynasty, and to more extent, during the Five Dynasties and Song dynasty, the Chinese landscape painting and the flower-bird painting had developed into maturity. Many landmark paintings and theories on the subject were generated in these periods. Since then, the theories and techniques of Chinese painting had influenced the development of Chinese garden. The “picturesque” quality in Chinese painting had affected the outcome of the design and construction of Chinese gardens in many ways. In Yuan, Ming and Qing dynasties, some painters were even contracted to create garden paintings. Among these paintings, the famous ones were the painting of the Garden of Lion Grove by Ni Zan of Yuan dynasty, the Garden of Eastern Mansion by Shen Zhou, the Eastern Garden by Wen Zhengming and the Zhi Garden by Zhang Hong of Ming dynasty, and the Forty Images of Yuan Ming Yuan or the Gardens of Perfect Brightness of Qing dynasty. These paintings were not only landscape paintings, but also the representations of the artistic conceptions of the gardens.

The emphasis of the Chinese culture and art is on the expression of the feelings and conceptions of human being. The art of Chinese garden is no exception. Ji Cheng said in his Preface in *Yuan Ye*, “I was known as a painter. I was by nature interested in seeking out the unusual. I

流芳园 – 晨曦

LFY - A morning view of the garden

苏州留园

The Lingering Garden in Suzhou

was fond of the brushes on the paintings of Guan Tong and Jing Hao, by which I was inspired all the time." The design and construction of garden should make it something like paintings and beyond the imaginations. Regarding the relationship between garden design and painting, Chen Congzhou said: "To seek the artistic conceptions and search for the charm of scenery to express lofty sentiments and meanings beyond the imagery. These two should be unified into one." "The picturesque images embody the poetic charm, while a garden is composed of the picturesque scenery. The quality of being poetic and picturesque is the essential guideline in Chinese garden design and construction." The Chinese landscape painting is closely associated with the art of Chinese garden. For example, the composition of a landscape painting is similar to the layout of a garden. The imagery of a landscape painting is similar to the scenery of a garden. As Shen Fu of Qing dynasty said: "The small scenery can be picturesque, the larger scenery can be ecstatic." Ji Cheng said: "To imagine beyond the confines of this secular world and feel as though you are wandering within a landscape painting." In short, the Chinese garden is a place where nature and human being is unified, a place where one can repose one's sentiment and conceptions in the landscape of the garden, and a place where one can refine the temperament and spirit.

中国艺术追求天然之趣，讲究气韵生动。恽寿平论画云："潇洒风流谓之韵，尽变奇穷谓之趣。"中国园林中的叠山理水，栽花种树，无不取法自然，又概括自然。园林中的范山模水、亭台楼阁和树石布置，都讲究参差不一，错落有致，虚实相生，以求"肇自然之性，成造化之功"。中国园林追求自然的古、拙、苍、野、逸等天然之趣。中国园林是浪漫的、幽雅的、自然的。"宛自天开"就是要呈现自然天成、气韵生动和神情轩朗的风采，要有生命活力和生命内涵。

"宛自天开"于中国园林中体现在多个方面，包括假山、水际、树木以及一些建筑材料。中国园林中的假山多由"人作"而成，大者如北京颐和园的万寿山，北海的团城岛，小者如北京故宫乾隆花园的假山，苏州环秀山庄的太湖石假山。这些假山无不具有天然真山之貌相、形态和质感。中国园林中的水，曲折多致，虚实相间，岸基起伏，烟霞无际，"静"者如镜，"动"者百态，无不表现出灵韵和生机。中国园林中的花木，姿态自然，竹木森秀，绿荫苍翠，四时清美，芳香四溢，无不孕育出天然气韵。中国园林中的建筑，石作、木作、瓦作、铺地，皆用自然材料；亭台楼阁，高低错落，古朴典雅；庭院曲折幽深，小中见大。凡此种种，无不与园林浑然一体。中国园林虽由人作而宛自天开，真有巧夺天工之妙！

Looking Like Naturally Created

The highest state of Chinese garden design and construction is as what Ji Cheng said: "though man-made, it should look like naturally created." Yang Xiong, a Western Han dynasty thinker, said: "Without the Heaven, man would have no reason to be human being; and without human being, the Heaven would not be completed its wholeness." The Chinese garden is a man-made environment aiming to reach the quality of nature. Zhuang Zi said: "The Heaven and the earth co-exist with me, and all things are unified with me as a whole." Human being and things are the integral parts of the universe. The "man-made" garden should follow the laws of nature, and at the same time, epitomize nature in an artistic way so that man and nature is unified into one.

流芳园 – 太湖石岸与睡莲

LFY - Taihu rockery bank and water lilies

The Chinese arts seek the charm of nature and the vitality of life. The composition of Chinese garden follows the course of nature. The mountain and water, pavilions and halls, rockery and plants, are all placed in natural ways as though they are created by nature. Some of the natural appearances such as the oldness, awkwardness, vigorousness, wildness,

苏州留园
The Lingering Garden in Suzhou

and elegancy, are all represented in the Chinese garden. For instance, the rockery mountain which is an important element in the Chinese garden is constructed by human being. But it embodies the essence of natural mountains in terms of profile, form and texture. Some of the best examples are the Mountain of Longevity in the Summer Palace Garden and the Tuancheng Island in Beihai Park in Beijing, and the rockery mountains in Qianlong Garden in Beijing and Mountain Villa Embracing Beauty in Suzhou. The water course in the Chinese garden is also constructed with natural forms such as the deep and hollow waters, winding banks of pond or lake, tranquil water and cascade water fall, etc. Even the buildings are constructed with many natural materials such as wood, stone, earth and mud tiles. All these present the ultimate goal of making a garden look like something naturally created.

瑶池蓬莱，琼楼玉宇 Wonderland Penglai, a beautiful house with a jade-like hall.

流芳园 – 玉茗堂

LFY - the Tea House: Hall of the Jade Camellia

第五篇

冶园笔意

Chapter Five

The Design of the Garden

意在笔先
Conceiving Before The Brush

中国艺术的创作讲“意在笔先”，“胸有成竹”或“胸有丘壑”。宋代书画鉴赏和画史评论家郭若虚在《图画见闻志》中曰：“意存笔先，笔周意内，画尽意在，像应神全。夫内自足，然后神闲意定。神闲意定，则思不竭而笔不困也。”“凡画，气韵本乎游心，神采生于用笔”。“意在笔先”，就是要有设计，要有思想。重要的是要明确创作的“意思”、“意义”和“意境”。以意运笔，需以心灵为指导，以生命之气势为重。《淮南子》曰：“故心者，形之主也；而神者，心之宝也。”计成《园冶·借景》云：“物情所逗，目寄心期，似意在笔先。”陈从周先生在《园林与山水画》一文中指出：“不知中国画理画论，难以言中国园林。” 造园即是如此，因心灵对场地的感悟而生意，以意设计，以意造园。“意在笔先”，按照现代的概念讲就是项目定位和规划设计。就造园而论，中心环节是设计，

In Chinese art and literature, the approach to creating an art work is to conceive before actually painting or writing. In Chinese terms, they are called *yi zai bi xian* or “conceiving before the brush”, “having a well-thought image of bamboo in mind” and “having a well-thought image of gully in mind”. Guo Ruoxu, a connoisseur and art critic of Song dynasty, said in his book of the *Record of Acknowledged Paintings*: “Conceiving before the brush, the meaning contains in the well organized brush strokes, the finished painting presents the meanings, and imagery conveys the essence of the subject. So after all these, I will be satisfied with it, and my spirit will be relaxed, my mind will be focused. In this atmosphere, my thoughts are inexhaustible and then my brush is free to paint.” “In any painting, the artistry comes from the mind, and the vivid spirit is conceived in the brush strokes.” This has explained the meaning of “conceiving before the brush” that is to have ideas and designs in advance. An artist should make it clear beforehand regarding the meaning, the significance as well as the artistic conception for the subject. It is the same in Chinese garden design. Ji Cheng said in *Yuan Ye*: “The attraction of natural objects, both the form perceptible to the eye and the essence which touches the heart, must be conceived in mind before putting brush to paper.” Chen Congzhou said in *The Garden and Landscape Painting*: “It is difficult to discuss the Chinese garden without the knowledge of the theories on Chinese painting.” It is true in the design of a Chinese garden. The artistic conceptions for a garden should be generated from the survey and characteristics of the site. Then, the design and construction of the garden should be based on these

其重点是要以“因地制宜”和“巧于因借”的原则来定位和设计园林的功能、布局和景致。

根据计成《园冶》中对相地的定义，流芳园这块场地基本符合“山林地”的特点，并且指出“园地惟山林最胜，有高有凹，有曲有深，有峻而悬，有平而坦……杂树参天，繁花覆地，自成天然之趣，不烦人事之工”。汉庭顿的大环境是一个人文荟萃和繁花似锦的宝地。这两点确定了流芳园应该是一座“城市山林的文人园林”，或者说它应该是一座具有“书卷气”的山水园林。

留存至今的文人园林主要在江南，集中在苏州，如苏州的沧浪亭，以及四大名园拙政

artistic conceptions. In contemporary terms, the notion of "conceiving before the brush" is to have project positioning and master planning and design before constructing a garden. The design phase is the central part of a garden project. It is a process of applying the principles such as "fitting in with the site" and "skillfulness on following and borrowing from" to the layout and design of the garden structures and scenes.

According to the definition in chapter of "Situating" in Ji Cheng's *Yuan Ye*, the site for the Huntington's Chinese garden — Liu Fang Yuan or the Garden of Flowing Fragrance, embodies the principal characteristics of the so-called "mountain forests". As Ji Cheng said: "The most

避暑山庄图（轴） 清　冷枚　故宫博物院藏
Leng Mei (1669-1742), The Mountain Resort, The Palace Museum

园、留园、网师园和环秀山庄等。另外，南京的瞻园、无锡的寄畅园，扬州的个园及何园，杭州的郭庄，上海的豫园等，也均为现存文人园林中之精品者。其中苏州园林与扬州的园林更具有秀气、文气、灵气。这些文人园林集中体现了中国园林艺术中对于追求自然精神、人格修养、心灵安顿和诗情画意以及“天人合一”的至高境界。

造园是以大地为素纸，绘出一幅园林世界。然而，造园无成法，园各有异宜，地与人俱有异宜，须善于用因。我在相地、构思和设计汉庭顿中国园——“流芳园”的过程中，不时记录下参悟所获之“意”和目营心构之“象”，园林的物象意趣渐渐清晰，从而达到“胸

picturesque site for a garden is among the mountain forests, where there are high and hollow terrains, winding and deep spaces, tall overhanging cliffs and flat level ground, all kinds of trees reach up to the sky, flowers cover the ground. The site has its own natural attractions without the touch of human handiwork." The Huntington, in a broader sense, is a precious place where the cultural richness is blended with the natural beauty of plants and gardens. These two prominent factors have determined that Liu Fang Yuan should be a so called "a scholar's garden in an urban forest". In another term, it should be a landscape garden with a "literary style".

盆菊图（局部） 明 沈周 辽宁省博物馆博物院藏

Shen Zhou (1427-1509), The Potted chrysanthemum (partial), Liao Ning Museum

中自有丘壑”、“意存笔先”和“心手合一”的园林设计状态。流芳园的总体立意是一座城市山水文人园，造园法则以计成的《园冶》和陈从周先生的《说园》为指导并以江南苏州园林为主要蓝本，并借鉴北方皇家园林的部分造园手法。其实，现存的明清时期的北京皇家园林如颐和园、圆明园和承德避暑山庄等，都有多处模仿江南园林的造园手法。陈从周先生曾提出："江南园林甲天下，苏州园林甲江南。”苏州园林又以文人山水园居多，以小中见大为胜，以隐逸林泉为趣，以淡雅精致为美。

因此，流芳园的设计总原则是借鉴苏州园林之胜，融合中国文人园林之精华，因地制宜地创造一座位于美国加州洛杉矶地区的传统式中国园林。流芳园的建筑风格以明清江南园林为主调，以典雅为其主要表现手法。石山以太湖石为主要材料，植物以本地植物为骨架，增植中国植物品种，与汉庭顿的植物园有机地融为一体。更重要的目的是为广大游客提供一个中国式的赋诗品园，畅聚交流，抚琴顾曲的“壶公天地”、“世外桃源”，也使流芳园成为当地华人的一处“精神家园”。

兴造园林，必从布局、构景开始，主要涉及空间结构、功能布置和景致安排。具体有建筑、理水、叠山、架桥、曲径和种植等方面，目的是达到“体宜因借”和“宛自天开”的艺术效果。先秦时期的《考工记》是中国最早的一部手工业技术和工艺美术文献，其中有曰："天

Most of the existing scholar gardens in China are located in the Jiangnan region, especially in the City of Suzhou, including the Cang Lang Pavilion Garden, the Humble Administrator's Garden, the Lingering Garden, the Master's Fish Net Garden and the Mountain Villa Garden in Suzhou, Zhan Garden in Nanjing, Ji Chang Garden in Wuxi, Ge Garden and He Garden in Yangzhou, Guo Villa in Hangzhou, and Yu Garden in Shanghai. The best scholar gardens are in Suzhou and Yangzhou due to their elegancy, literary and spiritual qualities.

The making of Chinese garden is like using the ground as a paper and drawing a painting of a garden world on it. There are, however, no fixed rules on the garden design and construction. Each garden should be different from one another because the site and environment as well as people are different in each case. All the differences should be taken into consideration and their advantages should be drawn on. During the process of site surveying, conceiving and designing the Huntington's Chinese Garden — Liu Fang Yuan, I recorded the detail information about the site and environment, wrote down my thoughts and made conceptual sketches for the garden. It was quite the same as the so called "conceiving before the brush" and "having a well-thought image of gully in mind". Therefore, I had a clear vision and an overall concept of the garden in the early stage of the project. Liu Fang Yuan should be a "scholar landscape garden sets in an urban forest". As for the theoretical references, those principles of garden design and construction presented in Ji Cheng's *Yuan Ye* and Chen Congzhou's *On Chinese Gardens* are used as the guidelines on Liu Fang Yuan

玩古图轴　明　杜堇　台北故宫博物院藏
Du Jin (1465-1509), Enjoying Antiquities, National Palace Museum

有时，地有气，材有美，工有巧。”它强调了造物的因时、因地、因材和因人的本质特征。造园亦是如此。在中国园林艺术中，水取其清澈婉转，石取其玲珑古朴，建筑取其空灵雅致，花木取其姿态优美和比德寓意，桥梁取其亲水通透，园径取其曲折幽深。园林布局，因地制宜，依山就水，巧妙布置，突出“景”字，抒发“情”字，传达“意”字，情景交融，诗情画意，则园林活矣！

project. As for its inspirational precursors in China, the gardens in Jiangnan especially in Suzhou are selected as the core models and some of the imperial gardens in northern China are models as well. Chen Congzhou once said: “Under the Heaven, the best gardens are in Jiangnan, while the Suzhou gardens are the best in Jiangnan.” Most gardens in Suzhou are scholar gardens best known for their qualities of “seeing large from the small” and their elegance and naturalness.

The general principle of the design for Liu Fang Yuan was to refer to the gardens of Suzhou and to integrate some of the best features of the scholar gardens in China. The focus of the design was to create a unique Chinese garden fitting into the particular site within the Huntington’s context and the climate of Southern California in the United States. The architectural style of Liu Fang Yuan referred to the style of Jiangnan gardens of Ming and Qing period. The rocks used in the garden were mainly Taihu rocks. The strategy for the vegetation or planting in the garden was to save most of the existing plants on the site and to add more Chinese plants and species. The ultimate goal of the project was to provide a place where visitors could, in a Chinese way or style, enjoy the garden, write poetry, play music and gather in a "land of idyllic beauty". In addition, I had wished that Liu Fang Yuan would become a spiritual home for the local Chinese communities as well.

The Chinese garden design and construction is a spatial art which integrates a variety of cultural, artistic and technical parts into a coherent whole. Those parts include piling up rocks, arranging water course, building structures, planting trees and flowers, decorating interiors, carving wood, making stone and brick couplets, composing lines of poetry, and drawing paintings inside a garden, etc., to create a meaningful, relaxing, elegant and close-to-nature living environment. The design of a garden should start from arranging spaces, functions and garden elements as well as sceneries. In *Kao Gong Ji* or *Record of All Crafts*, a book of Spring and Autumn Period, it said: "The Heaven holds the time, the earth has Qi or energy, the material has its beauty, the craft has its artistry." It underlines the essential point that to create a thing should be in accordance with the characteristics of the time, place, material and human being. This notion is true in designing and constructing a garden. The art of Chinese garden design and construction, for instance, is to create and assemble the attractive and artistic features of the garden such as the transparency and mellowness of water, the exquisiteness and quaintness of rocks, the gracefulness and elegancy of architecture, the beautiful stance and metaphorical meanings of plants, the closeness to water and openness of bridges, and the winding and depth of paths. A garden should be laid out skillfully to fit in with the site, and it sit close to mountain and water, to give prominence to the scenery, to express the sentiment, and to convey the artistic conception. The integration of the poetic sentiments with picturesque imageries will make a garden vivid and colorful.

流芳园 － 建筑细部
LFY - building detail

建筑庭院

中国园林是生活空间的延续和一种生活方式的体现，它必然要有构成生活方式的内容和文化；既然是一处精神家园，它就应该反映人的哲学思想和生命理念。中国园林建筑及庭院是构成中国园林活动空间的最主要元素。中国最古老的风水经典《黄帝宅经》曰："宅者人之本。人因宅而立，宅因人得存"，"人宅相扶，感通天地"。建筑是立于天地之间的人居之所，是安身立命的空间。中国园林建筑是兼具功能性、娱乐性、艺术性和精神性的"宅"。计成在《园冶·立基》中就云："凡园圃之基，定厅堂为主"。《园冶》中有近五分之三的篇幅论述园林建筑，足以说明园林建筑在中国传统园林中的重要性。"楼阁亭宇，乃山水之眉目也"（清郑绩《梦幻居画简明》）。陈从周先生说："一个园林里有建筑物，这就有了生活。有生活才有感情，有了感情，这才有诗情画意。"由此可以看出，建筑在中国园林中有重要地位，也可以说中国园林不可能脱离建筑而存在。"园林是建筑的延伸和扩大，是建筑进一步和自然环境（山水、花木）的艺术综合，而建筑本身，则可说是园林的起点和中心"（金学智《中国园林美学·建筑之美》）。在中国古代，没有建筑的园林，只能谓之"林"，不能谓之为"园"。西汉方士公孙卿曰："仙人好楼居。"

Buildings and Courtyards

The Chinese garden is an extension of living space and a reflection of life style. Buildings and courtyards are the actual functional spaces for the activities taking place inside a garden. In one of the most ancient treaties on *feng shui* or wind and water, *Huang Di Nei Jing* or the *Inner Canon of Huangdi* said: "House is the foundation of man. A man can live well because of the house, while the house can be useful because of the man." "House and man sustain each other, so Heaven and earth are sensible by man." Architecture or building is an abode for people to live in between the Heaven and the Earth. Buildings inside a Chinese garden are the "dwelling houses" for functional, entertainment and artistic uses and as a spiritual "abode". Ji Cheng said in *Yuan Ye*: "The most important element in the layout of a garden is to situate the principal buildings." In the book, Ji Cheng wrote almost three-fifths of the content on buildings and structures of gardens. This means that buildings were very important elements in traditional Chinese gardens. "Towers and pavilions are the eyes and eyebrows of the mountains and water in the landscape," said Zheng Ji (AD 1813—1874) a scholar of Qing dynasty. Chen Congzhou said: "There are buildings inside a garden, so that there are lives and activities happening in the garden where the sentiments are generated, then the poetic and artistic conceptions are stimulated." It is said that the Chinese garden would not exist without buildings. In his book of the *Aesthetics of Chinese Gardens*, Jin Xuezhi said: "Garden space

中国古代传说中的“象天建筑”，如观象台、灵台和明堂等，也是中国园林建筑的渊源之一。

中国园林的建筑有厅堂、楼阁、亭台、船舫、轩馆、书斋和廊榭等部分。明代锺惺在《梅花墅记》中概括了四种园林建筑的特性，曰：“高者为台，深者为室，虚者为亭，曲者为廊。”古时还有“堂以宴、亭以憩、阁以眺”的造园法则。中国传统建筑以木结构为主，从建筑形式方面讲，一座建筑主要由三部分组成，即台基（高）、屋身（边）和屋顶（上）。各部分功能作用有所不同，按照《墨子·辞过》所云：“为宫室之法，曰：室高，足以辟润湿；边，足以圉风寒；上，足以待霜雪雨露。”从形式美的角度来看，下部“台基”主要有建在陆

is an extension of the architectural space. It integrates the building with natural environment such as trees and flowers. Building becomes the anchor and the central element in a garden." In ancient China, a garden without building would only be called "a forest", not a "garden". Tracing the source back to the ancient times, the Chinese alchemists such as Gongsun Qing of Western Han dynasty (206 BC—AD 25) said: "The Immortals liked to live in a tower building." Another source of the garden buildings came from the tall terraces and buildings in the landscape for ritual activities such as observing and worshiping the Heaven and earth. All these are the original reasons of having many buildings inside gardens.

圆明园四十景之濂溪乐处

Yuan Ming Yuan, one of the Forty Views - A Pleasant Place by a Stream

地上的石座台基或架在水上的柱式台基，两者都有极好的稳定感和厚重感，与轻巧、空灵的栏杆结合形成优美的基座以烘托上部主体建筑的雄浑。中部“屋身”主要由墙体、门窗和立柱组成，外观挺拔、端庄。江南建筑更显得轻盈疏透和精致典雅。上部“屋顶”主要由屋面、屋檐和屋脊构成，造型多样，总体外观显得飘逸、飞动和舒展。整体建筑虚实相生、轻盈流畅，与山水园林环境融为一体。园林建筑布局以规则为主、朝南为主、围合为主，少数建筑结合地形和借景需要而自由布置。园林建筑及其庭院的围合使园林空间层次丰富、曲折有致、小中见大，形成“园中有园，景外有景”。

中国传统住宅建筑包括园林中的主体建筑绝大部分是以南北向规则布置，它们更具功能性、实用性，它们的表象是“入世”的、礼制的、人为的。住宅建筑讲等级，园林建筑论主次。而园林山水则以自由方式布置，得景随形，它们的表象是“出世”的，浪漫的、自然的。中国园林中融合了“规则与自由”、“现实与浪漫”、“人工与自然”，反映了中国人的“天人合一”的思想。冯友兰先生在其《中国哲学简史》中说：“中国哲学‘既入世而又出世’，中国哲学并不把‘人’与‘天’对立，而是人与自然并生。”中国人讲究“入世”与“出世”的平衡，即所谓“不离日用常行内，直到先天未画前”。由此可见，中国园林中的建筑庭院与山水花木形成了一种有趣的“对比”与“并列”的关系，是一种

There are various types of buildings in the Chinese garden, including the great halls (*ting tang*), towers (*lou ge*), pavilions and terraces (*ting tai*), boat buildings (*chuan fang*), houses (*xuan guan*), studies (*shu zhai*), covered walkways (*lang xie*) and bridges (*qiao liang*). Zhong Xing (AD 1574—1624), a scholar of Ming dynasty, said in the Record of Plum Flower Villa: “The tall terrace is *Tai*, the deep structure is *Shi*, the empty structure is *Ting*, and the winding structure is *Lang*.” In the old times, each building was dedicated to a specific function, such as a hall for banquet, a pavilion for relaxation, and a tower for gazing afar. Most of the traditional Chinese buildings were wooden structures. A building is composed of three parts: the foundation platform, the middle wooden structure and the decorative roof. The original function of each part was clearly defined: the foundation platform is for protection from the moisture of the ground, the middle structure of walls for protection from the wind, and the roof for protection from the rain and snow. Each part of the building has its own aesthetic appearance: the foundation platform with a sense of steady and heavy, the middle structure forceful, decorative and beautiful, and the roof upward, flying and elegant. These features of garden buildings are well blended with other landscape elements such as hills, ponds, rocks and plants. The layout of buildings is usually in an orderly fashion, where most of the buildings are oriented in the north-south direction, and clustered into courtyards. In this way, a garden has layers of spaces, twists and turns, and a variety of scenes.

弘历行乐图（局部） 清 张廷彦 故宫博物院藏
Zhang Tingyan (1735-1794), Hong Li Traveling for Pleasure, The Palace Museum

物质享受与精神追求的并列，是现实主义与理想主义的并列。所以，从总体上来讲，中国园林建筑和庭院从来不是无“章法”的自由布置，而是以一种“规则”与“自然”相并列和融合的方式布置。这一点可以从历史上的园林绘画和现存的园林布局中得到验证。

中国传统建筑特别是园林建筑，主要以群体组合方式布置，十分讲究运用“虚”与“实”、“露”与“藏”、“有限”与“无限”以及“巧于因借”的造园手法，使园林具有“空灵”的境界。计成在《园冶·屋宇》中对园林建筑的布置有精彩的描述：“奇亭巧榭，构分红紫之丛；层阁重楼，回出云霄之上；隐现无穷之态，招摇不尽之春。” 近代古建筑学家朱

Most of the scholar gardens in China were private gardens. They were directly connected with the dwellings, and were actually part of the living quarter of a residential complex. They were secluded places within the urban environment so that the owners of the gardens could enjoy the convenience of daily life in the city while strolling inside the gardens to enjoy the landscape. Although not like the layout of imperial palaces or temples, which are usually symmetrical and axial to express the beauty of solemnity, the main buildings inside a Chinese garden are laid out in some kind of order too. From the philosophical point of view, a building complex is a presentation

启铃在《园冶注释》序中说："吾国建筑，喜用均齐之格局，以表庄重……若夫助心意之发舒，极观览之变化，人情所喜，往往轶出于整齐画一之外……纯任天然，可以尽错综之美，穷技巧之变……"在园林中，建筑与游廊结合，围合成大小不一的庭院或"园中园"的空间意境，以化大为小、化集中为分散的手法将建筑与山水、花木组成园林景观，形成风景中的建筑与建筑中的风景相映成趣的园林景观。游廊是中国园林的特色之一，在园林建筑中占据重要地位。游廊不仅具有连接建筑庭院和分隔空间的功能，而且提供引导和观景的作用。中国园林中的游廊形式也有许多，如横跨池沼的水廊，愈折愈曲的曲廊，随势起伏

of "secularism" or worldly life which the Confucius wanted to pursue, while the landscape of a garden is a presentation of the "spirituality" or non-worldly life which the Daoism wanted to pursue. The former reflects the realism and the beauty of order, while the later the idealism and the beauty of nature. Most buildings of the living complex and major buildings in a garden are oriented in the north-south direction in a hierarchy order, while the pavilions, covered walkways and bridges, etc., are organized in accordance with the characteristics of the site. In the Chinese garden, there are juxtapositions of "order and freedom", "realism and idealism", "artificiality and nature" as

苏州拙政园

The Humble Administrator's Garden in Suzhou

流芳园 – 建筑飞檐
LFY - The wingning roof of a pavilion

well as the material comfort and spiritual enjoyment. In fact, they are the reflections of the Chinese ideology of the "unity of Heaven and human being". Therefore, the layout of most buildings and courtyards inside a Chinese garden is never randomly organized or "freely" arranged. It is a juxtaposition and integration of buildings and the garden landscape elements such as mountains, waters and plants. This phenomenon can be found from the historical garden paintings and the existing Chinese gardens as well.

The traditional Chinese architecture, especially the garden architecture, was organized in clusters of buildings and courtyards. The spaces are enchanting and features are unique with the juxtaposition of emptiness and fullness, exposition and seclusion, limit and infinity in spaces. Ji Cheng said in *Yuan Ye* on buildings: "Unique pavilions and fine gazebos should be built separately among red and purple flowers; storied towers and lofty buildings should tower above the clouds; different prospects will appear and disappear inexhaustibly; endless spring scenery will hover over your garden." The integration of buildings and covered walkways in Chinese garden will create many enclosed spaces in different scales so as to turn a large complex into a series of small garden spaces. The hills, waters and plants are assembled to create attractive scenes. Covered walkway is a unique Chinese garden feature. It plays an important role in making a Chinese garden. The covered walkways serve not only to connect buildings and divide garden spaces, but also to guide visitors to the scenery. There are many different types of covered walkways such as the ones crossing ponds,

的爬山廊等。游廊使园林空间产生小中见大、虚实相生和迂回幽深的效果。中国园林中的开敞式建筑，如亭、廊、榭等，与园林中的湖、塘、池、潭等水体，都是园林中的“空”。这些“空”的空间，正似中国画中的“留白”。令人遐想，令人顿悟。朱良志先生说，“空则可纳万象，可得天全。空的境界才是自由的”。时代在变化，但山水精神没有变，正如禅宗所妙悟的那样：“万古长空，一朝风月”，这也是造园的最高境界。

中国园林建筑的风格在三千余年的历史当中变化并不大。中国艺术追求的是生命精神，一种超越时空的永恒，即所谓“荣落在四时之外”的“象外之象”。中国园林建筑风格无论是秦汉时期的朴拙雄壮之美，唐宋时期的恢宏俊秀之美，还是明清时期的典雅细腻之美，它们所体现出的都是中国人对人生、对自然的理想，对山水风月的情怀。中国传统建筑的

winding through leveled areas, and climbing up to hills, etc. The covered walkway will create an effect of seeing the large through the small, growing out of emptiness and fullness, and winding and deep spatial atmosphere. The open structures such as pavilions, covered walkways and gazebos, similar to the water body of pond or lake, are the void or empty spaces in a garden. The emptiness of the space can accommodate various scenes around. It is similar to the blank area in a Chinese painting, which provides imaginable space for various things. “The prospect of emptiness is the freedom”, said Zhu Liangzhi.

苏州狮子林

The Loins Grove Garden in Suzhou

流芳园 - 庭院空间

LFY - Courtyard space

历史延绵数千年，虽有南北地域风格的差异，南方园林以小巧细秀和淡雅素朴之秀美为主，北方宫苑以高大壮观和富丽堂皇之华美为主，但总体风格变化不大。按照计成《园冶·屋宇》中表达出的造园思想，建筑应该“时遵雅朴，古摘端方”。

流芳园里的主要园林建筑和庭院融合了中国南北两地的空间格局，以规则式和南北朝向为主。按此原理布置，布局空间甚至还有局部“轴线”和“对景”的关系。其他一些廊、桥、亭等则随地形布置。主体建筑庭院，游廊环绕，空间错落有致，直中有曲，实中有虚，小中见大，大中见小，景致各异。园林建筑风格以现存于世的明清风格为蓝本，以苏州园林为主要参照。在中国园林当中，苏州园林是最具文人气质的山水园林，苏州也是现存文人山水园最集中的地区。所以，在汉庭顿这样一个人文气息十分浓郁的环境里建造一座中国传统式文人山水园林，以苏州园林为主要借鉴是十分合适的。

The style of the Chinese garden architecture had not changed dramatically in the past three thousand years. The spirit of life is the essence of the Chinese art. The Chinese art represents the eternity beyond time and space, “the exuberance and decadence beyond the four seasons”, “the appearance beyond imagery”. Regarding the architecture of Chinese gardens, no matter whether it is austere and magnificent in Qin and Han dynasties, or spectacular in Tang dynasty, or exquisite and elegant in Ming and Qing dynasties, presented the Chinese ideals about human life and nature, and the sentiments toward the landscapes. In the past millennia, the traditional architecture of Chinese gardens, though different in the Southern and the Northern regions, is consistent. Therefore, the most important principle in garden architecture is to “follow what is elegant and simple at the time, and take what is the most orthodox from ancient times”, said Ji Cheng in *Yuan Ye*.

The spatial organization of the buildings and courtyards in Liu Fang Yuan is an integration of the southern and northern gardens in China. Most of the main buildings and courtyards are arranged in an orderly fashion and facing the south. Occasionally, the layout of axes and look-straight views are applied. The pavilion, bridges and covered walkways are organized in accordance with the lie of the site. The architectural style is similar to the style of Ming and Qing dynasties. Most of the buildings refer to those in Suzhou gardens. Among the Chinese gardens, most scholar landscape gardens are located in Suzhou. Therefore, it is highly appropriate to select the Suzhou gardens as the references for the Huntington’s Chinese Garden because the Huntington institution is a place with literary quality and atmosphere.

理水架桥

“名园依绿水”，这是中国园林的基本特性。清代画家笪重光在其《画筌》中曰：“山脉之通，按其水径；水道之达，理其山形”；“水为先者，意中有水，方许作山。” 陈从周先生说：“园必隔，水必曲”；“水随山转，山因水活。”水的审美特性有“洁”、“虚”、“动”、“纹”。所谓“天下之山，得水而悦；天下之水，得山而止”（明归有光《宝界山居记》）。

水在中国园林中占据重要位置。《周易・说卦传》曰：“说万物者莫说乎泽，润万物者莫润乎水。终万物始万物者莫盛乎艮。”西晋文学家左思《招隐》曰：“非必丝与竹，山水有清音。”明代文震亨在《长物志》一书中曰：“石令人古，水令人远，园林水石，最不可无。”又曰：“石令人幽静，水令人旷达。”中国山水画中常以水作为画中生机之源，“山以水为血脉”。中国园林更有“无水不成园”之说。园林有水，有水则有空灵，有空灵则园林活矣。中国古代诗人写过许多千古传诵的咏水名句，如“云日相晖映，空水共澄鲜”（南朝宋谢灵运《登江中孤屿》），“野旷天低树，江清月近人”（唐孟浩然《宿建德江》），“明月松间照，清泉石上流”（唐王维《山居秋暝》）。陈从周先生说过：“‘空灵’二字，为造园之要谛”，“池水无色，而色最丰”。园林之水，若静若动，若有若无，影落水中，

Arrangement of Waters and Bridges

One of the fundamental identities of Chinese garden is that “the best gardens are built in accordance with green waters”. Da Chongguang (AD 1623—1692), a Qing dynasty painter, said in *Hua Quan* or the *Annotation on Paintings*: “The openness of mountains follows the waterways; where a waterway reaches, the shape of the mountain is formed accordingly.” “The water element should be conceptualized first, then the mountain.” Chen Congzhou said: “A garden should be partitioned, the water be crooked.” “Water flow follows the course of mountains, while mountains become alive because of the water.” The aesthetic characters of water include cleanness, transparency, motion and veins. Gui Youguang (AD 1507—1571), a Ming dynasty writer, said in the *Record of Bao Jie Mountain Dwelling*: “Mountains under the Heaven are delightful because of the water around them, while waters under the Heaven are stopped because of the mountains.”

Water is an important element in the Chinese garden. In *Zhou Yi*, it said: “Nothing is more joyful than the lake in pleasing all things, nothing is more humid than water in moistening all things, and nothing is more exuberant than the mountain in completing their cycles and re-generating of all things.” Zuo Si (AD 250—305), a Western Jin dynasty writer, said in *Zhao Yin*: “There is no need for string and bamboo instrument, as mountain and water make mellow sounds.” Wen Zhenheng (AD 1585—1645), a Ming dynasty scholar, said in *Chang Wu Zhi* or *On Superfluous*

圆明园四十景之曲院风荷
Yuan Ming Yuan, one of the Forty Views - The Courtyard and Lotus Breeze

虚实相应，一片灵韵。所谓“盈盈一水间，脉脉不得语。”中国园林中的理水讲究曲折变化，如宋代画家郭熙论山水画曰：“山欲高，尽出之则不高，烟霞锁其腰则高矣。水欲远，尽出之则不远，掩映断其流则远矣。”

在中国文人山水园林中，“理水”在总体布置中占有极其重要的位置。古人有“假山可为，假水不可为”之说。“理水”首先是要探寻利用自然水源，其次是水形和水景的布置。中国园林中的水取其清澈曲折和自然风貌为主。水面曲折而产生流域广阔，望之不尽和小中显大的效果。通常水面被桥、堤和岛屿隔为大小不同的空间，有开阔的中心水面，有水湾、

Things: “Stone makes a person feel ancient and peaceful, and water makes a person feel distant and open-minded; so stone and water is the most indispensable elements in a garden”. To the Chinese, if there is no water in a garden, it would not be called a garden. Water produces an effect of openness and spaciousness, which makes a garden alive. In Chinese poetry, there are many wonderful verses to describe waters. Xie Lingyun, a Southern dynasty poet, wrote: “Clouds and the sun shine on each other; open water is clear and fresh.” Meng Haoran, a Tang dynasty poet, wrote: “The horizon of a wild open field is lower than the trees; the clear water in a river reflects the moon closer to people.” Wang Wei, a Tang dynasty poet, wrote: “The bright moon shines between pine trees; clear spring

渊潭、溪涧等。水岸则模仿自然形态散置叠石。唐代诗人白居易在描述自家园池时，总结出理水的一个原则：“岸浅，桥平，池面宽。”

流芳园的水景布置采用了传统的手法，以一汪水际作为整个园林的中心，以桥和堤将水面划分成大小不一的四个水空间，中间者为最大，东面者次之，西面者再次之，西北面者最小。利用地形特点，理水以聚为主、聚分结合，将主要水体用平桥、曲桥、拱桥和岛屿分隔成中心湖泊、水塘、水湾、渊潭四个水面。再结合场地南北两个方向的天然溪流，将园中的水景组织成池、潭、湖、涧、瀑、泉等多种形态，使整个流芳园的水景在空间上和形态上都十分丰富，基本上涵盖了中国文人山水园中水景的各种特征。在季节时令变化中，会出现月夕澄波、雨后霞影、晨起青烟、水流云在等丰富的园林景观。

中国园林中必有水，有水就有桥梁。无论是一条石梁跨水而过，一曲平桥连接两岸，一轮“月”拱桥倒映水中，还是数孔石桥横卧水上，这些都已成为中国园林中的别致景观。唐代诗人杜牧写扬州之景：“二十四桥明月夜，玉人何处教吹箫。”王勃有诗曰：“泉声喧后涧，虹影照前桥。”古时中国人称桥为梁，《诗经·大雅》：“维鹈在梁，不濡其翼”。到了汉代，梁改称为桥。宋之前，以木桥为多，宋之后，石桥居多。

tumbles through rocks.” Chen Congzhou said: “*Kong ling* or openness and spiritual are the gist of garden design and construction.” “Water in a pond looks colorless, but it reflects rich colors.” Waters in a garden, tranquil or moving, real or illusive, or reflections, create vitality and rhythm. In the Chinese garden, waters are arranged with twists and turns.

The arrangement of waters inside a Chinese garden is extremely important in the layout of a garden. The ancient Chinese said: “Man-made hillocks can be constructed, but man-made water is not possible.” To arrange water features, the first task is to find the natural source of water, the second is to make the pattern of the water, and the third is to lay out the water scenery. The water scenery in the Chinese garden is utilized for its clearness and natural appearance of twists and turns which create the effects of openness, endlessness and enlargement. In most cases, waters in a garden are divided by small bridges, causeways or islands into central pond, bay, deep pool or stream. The bank of a pond or lake is usually built with scattered rocks with a very naturalistic manner. Bai Juyi, a Tang dynasty poet, summarized the principle of the water scenery arrangement: “Lower banks, level bridges, and a wide open pond surface.”

The layout of the water scenery in Liu Fang Yuan is designed in accordance with traditional principles. A puddle of water or the lake becomes the center of the garden. The lake is divided into four differently sized water spaces, of which the central one is the largest, the eastern smaller, the western even smaller, and the northwestern the smallest. All these water scenes are assembled

"桥"是中国文人的一种情思！陈从周先生说："桥给人以画一般的意境，诗一般的情感。"中国园林中的桥是构成园中的景点乃至景观标志之一，如北京颐和园的"十七孔桥"，杭州西湖的"断桥残雪"，扬州瘦西湖的"二十四桥"，苏州拙政园的"小飞虹"廊桥等。中国园林中的桥梁根据园林的大小和水面的宽窄，设置不同尺度和样式的桥。在流芳园里，布置有各种式样的石桥数座，有石板桥、石拱桥、一孔桥、三孔桥、曲桥和廊桥等。桥的比例适度，平缓易行，造型凝重，色调拙朴，与周围垂柳、倒影、亭阁和谐一致。中国人爱桥，多地有"走桥"的风俗，以图吉利、平安和长寿，还有赏桥看月的习俗。长桥映月，

around the central area of the garden. And at the same time, they are divided by bridges and islands to create different water views of lake, pond, bay, deep pool and stream. These various water scenes are typical in Chinese gardens. They have enriched the landscape scenery of the garden. In different seasons and times, various scenes in the garden will appear: the reflection of the moon on the clear waves, the shadow of clouds after rain, the mist in the morning and the tumbling water merge into the upper clouds.

苏州拙政园
The Humble Administrator's Garden in Suzhou

画舫弦歌，湖亭美酒，一桥相望。这些“桥文化”已经融入中国人的生活当中。园林之桥，更是与诗情画意联系在一起，成为中国园林中不可或缺的组成部分。

Water is an inevitable element in Chinese gardens. Wherever there is water, there are bridges. A stone slab stretches across the water, a leveled winding bridge connects the two sides of water, a arched "moon" bridge is reflected in the water, or a multi-arched stone bridge reclines on the water. All these are the unique scenery of bridges in Chinese gardens. Du Mu (AD 803—852), a poet of Tang dynasty, described the scenery of Yangzhou: "The clear moon hangs over the twenty four bridges; where is the noble scholar teaching flute music?" Wang Bo (AD 650—675) of Tang dynasty wrote: "The sound of the spring rackets the stream in the back; the reflection of rainbow sets on the bridge at the front." In ancient China, most bridges were built with wood until Son dynasty, when stone bridges were popular in most cities and gardens. "Bridge" embodies the emotions of the Chinese. Chen Congzhou said: "Bridges bring artistic conceptions and poetic sentiments." Bridges in some of the Chinese gardens become landmark features that make those gardens well-known and distinguished from one another, such as the Seventeen Arches Bridge in the Summer Palace in Beijing, the Lingering Snow on the Broken Bridge in Hangzhou West Lake, the Twenty Four Bridges in the Slender West Lake in Yangzhou, and the Little Flying Rainbow Bridge in the Humble Administrator's Garden in Suzhou. Bridges are different in accordance with

流芳园 – 步月石拱桥
LFY - Bridge of Strolling in the Moonlight

扬州何园 - 石桥
He Garden in Yangzhou - A stone bridge

the size of the garden and water. Inside Liu Fang Yuan, about ten different bridges are designed or built, such as stone slab bridge, stone arched bridges, covered walkway bridge, etc. These bridges are in proper scales and proportions with gentle slopes, elegant styles with plain color. They are blended well with the pavilions, willow trees and reflections in the water. In many areas of China, "to cross a bridge" becomes an old custom which is believed to be able to bring good fortune, good health, longevity and even prosperity to those who cross the bridge. Bridges are also good spots for watching the moon. Therefore, the garden bridge is an indispensable part of the Chinese garden.

叠山掇石

“叠山理水”是中国园林造园之根基。其中“叠山”又分为“叠石”和“掇山”两部分，叠石重点在以石点景，掇山侧重以石造山。中国人有玩石、赏石的传统。古人认为，“石不能言最可人”，“人以石清，石以人贵”。又有“石令人古”之说。中国的石文化悠久而深邃，它反映了中国人的人生哲学、审美情趣和精神境界。古人曰：“一石清供，千秋如对。一拳之石，而能蕴千年之秀。”又有“山无石不奇，水无石不清，园无石不秀，室无石不雅”之说。

中国古代文人墨客喜好以石为友，以石交友。最知名的有唐代诗人白居易爱太湖石，“待之如宾主，视之如贤哲，重之如宝玉，爱之如儿孙”。他在《太湖石记》一文中描述了对石的痴迷：“古之达人，皆有所嗜。玄晏先生嗜书，嵇中散嗜琴，靖节先生嗜酒，今丞相奇章公嗜石。石无文无声，无臭无味，与三物不同，而公嗜之，何也？众皆怪之，我独知之。三山五岳、百洞千壑，覙缕簇缩，尽在其中。百仞一拳，千里一瞬，坐而得之。此其所以为公适意之用也。”他还认为太湖石是“岂伊造物者，独能知我心”。宋代书画家米芾酷爱奇石，达到了顶礼膜拜的程度，故有“米芾拜石”的典故。宋代文学家苏轼也爱奇石，

The Piling-up of Rocks and Hillocks

“Piling up hillocks and arranging waters” are essential to construct Chinese gardens. “Piling up hillocks” has two components: one is to pile up peak rocks and the other hillocks or mountains. “Piling up rocks” is to place peak rocks to decorate key spots in the garden, while “piling up hillocks” is to use rocks to create mountain-like hills. The Chinese have a tradition of appreciation of rocks, believing that “stones are speechless, but lovely”, “man becomes distinct like stone, while stone becomes noble like man”, and “stone makes a person feel pristine”. It reflects the cultural and philosophical sentiments of the Chinese. The Chinese also believe that “looking at a displayed stone is like seeing through a thousand-year history. A stone as large as a fist embodies the splendor of a millennium”, and “without stones, mountains are not wonderful, water not clear, gardens not beautiful, rooms not elegant”.

The ancient Chinese scholars were fond of making friends with stones and making friends through stones. Bai Juyi, a prominent poet of Tang dynasty, was obsessed with Taihu rocks. He treated them as guests, revered them as sages, valued them as precious jades, and loved them as his children. He also believed that the Taihu rock was “only the created being who could understand me”. Mi Fu (AD 1051—1171), a Song dynasty painter, was extremely fond of stones. He had even worshiped stones. Su Shi (AD 1037—1101), a prominent writer of Song dynasty, was also a

扬州瘦西湖 － 太湖石
The Slender West Lake in Yangzhou - Taihu rock

他说“园无石不秀，斋无石不雅。”宋代徽宗皇帝造“艮岳”园林，采集天下奇石于一园，而有运送奇花异石的“花石纲”之奇举，使艮岳“丘壑林塘，杰若画本，凡天下之美，古今之胜在焉”（宋代张淏《艮岳记》）。清代画家石涛也是一位爱石和叠石掇山之人。据传保存至今的扬州何园内的“片石山房”就是他的遗作。石涛曰：“山林有最胜之境，须最胜之人，境有相当，石我之石，非我则不古；泉我之泉，非我则不幽。”清代书画家郑板桥以画竹石最著称，他也酷爱顽石，并曰：“得美石难，得顽石尤难，由美石转入顽石更难。美于中，顽于外，藏野人之庐，不入富贵之门也。”中国的赏石文化不仅给人以心情愉悦，而且通过一方顽石而能体味生命的价值。

中国文化中的赏石，也称为玩石、供石。中国四大名石包括太湖石、灵璧石、英石和昆石。体量较大者可作为厅堂石或庭院石来赏玩，体量较小者则适合摆放在桌案上和把玩。中国园林当中的赏石多为太湖峰石，置于园中或庭院里。而园林建

流芳园 － 假山、石桥
LFY - The Rockery Hillock and a stone bridge

stone lover. He said: “Gardens are not beautiful without stones, and houses not elegant without stones.” The Song emperor Huizong (AD 1082—1135) built a magnificent garden called Gen Yue. He ordered the collection of abundant rare rocks and exquisite plants from all over China putting them inside the garden. This undertaking was known as “Hua Shi Gang”, which made the garden an ensemble of picturesque hills and waters of the world and of the past. Shi Tao (AD 1630—1724), a Qing dynasty painter, was a stone lover as well as a garden master. The existing rockery in Pian Shi Shan Fang or Mountain House in Yangzhou was attributed to Shi Tao’s work. Zheng Bangqiao, a Qing dynasty painter, was a stone lover as well and famous for painting bamboo and rocks. The Chinese culture of appreciation of stones gives people not only pleasure but also appreciation of the value of life.

筑里面则常常在厅堂摆放赏石或供石。中国人欣赏奇石，喜欢它的“瘦、漏、透、皱”，也欣赏它的“清、顽、丑、拙”。“瘦”，挺劲修长的风骨；“漏”，血脉畅通的活力；“透”，玲珑剔透的意态；“皱”，肌理变化的韵味。“清”，清秀之美；“顽”，野逸之气；“丑”，奇突之态；“拙”，浑朴之质。奇石可谓大千世界的缩影，一石一景，一石一物，一石一天地，一石一世界。中国人爱石，不仅爱其形、其质，而且通过石来品味人生、抚慰生命。中国画家也喜爱画秀石，表达画家的心中之境和无言之美。元人许有壬有诗云：“两竿瘦竹一片石，中有古今无尽诗。”明人计成云：“片山块石，似有野致。”现存于中国园林当中最著名的四大太湖峰石是苏州留园里的“冠云峰”和“岫云峰”，苏州十中校园里的“瑞云峰”和上海豫园里的“玉玲珑”。园林若得上品奇石，可使庭院蓬荜生辉，园林名扬天下。

中国人欣赏石，其实是从欣赏山开始的。先秦古籍《山海经》一书就有五篇记述中国大江南北的诸山，如：“岷山，江水出焉，其下多珉。”孔子曰：“知者乐水，仁者乐山。” 上古时期有帝王登山和祭山，秦汉时期有园林中设“三山”以求仙，魏晋时期有士大夫隐居山林，唐宋元时期有山水诗文绘画，明清时期有文人山水园林。山，已融入中国人的生活和情感之中；山，已有了中国人的人格品质和追求的象征。清代画家戴熙《习苦斋题画》曰：“春山宜游，夏山宜看，

苏州留园 – “冠云峰”
The Lingering Garden in Suzhou - the Guan Yun Peak Rock

The most beloved stones in China are Taihu rock, Lingbi stone, Ying stone and Kun stone. The large-sized stones are usually placed inside gardens, the middle-sized ones in courtyards, and the small-sized ones on tables inside halls or pavilions. In Chinese gardens, Taihu stones are mostly used. The Chinese appreciates the exquisite formation and texture of these kinds of stones, namely *shou*, *lou*, *tou* and *zhou*, or slenderness, pierced, openness and wrinkles of a stone or rock. Each of these four shows different characters in stones: *shou* shows the strength, *lou* the vitality, *tou* the artistry, and *zhou* the rhythm. Stone is a common element in Chinese paintings as well. Xu Youren (AD 1286—1364), a Yuan dynasty poet, wrote: “Two stems of bamboo and a piece of rock contain countless poetry from ancient to modern times.” Ji Cheng said: “A part of mountain or a piece of rockery resembles wildness.” The most well-known four Taihu rocks existing today are Guan Yun Feng and Xiu Yun Feng in Suzhou Lingering Garden, Rui Yun Feng in Suzhou city, and Yu Ling

秋山宜登，冬山宜居。”

中国园林以山水为骨架，无水不成园，无山则不为园。在中国传统园林中，堆土、叠石，被称为造假山，是中国造园艺术中极为重要的一个组成部分。造假山这门艺术，距今已有两千余年历史，《尚书》中就有“为山九仞，功亏一篑”的记载。秦汉时期所建造的建章宫中的“太液池”和“方丈、蓬莱、瀛洲”三座人造神山在《史记》中有明确记载。之后，历代都有堆土叠石的假山杰作，其中宋徽宗的艮岳大假山最为知名（现已不复存在）。江南园林在明清时期盛行以石叠山，石材主要有太湖石和黄石。假山已成为园林之形胜的代表。计成曰：“池上理山，园中第一胜也。”中国园林假山经历了从以堆土为主，到土石结合为主，再到以叠石为主的发展过程。计成在《园冶》中则称：“雅从兼于半土”，认为半土半石的假山品位最为高雅。《园冶》在“掇山”和“选石”两篇中专门论述了叠山的原理和手法。计成论述掇山之要义时云：“察乎虚实”，“瘦漏生奇，玲珑安巧”，“岩、峦、洞、穴之莫穷，涧、壑、坡、矶之俨是”，“蹊径盘且长，峰峦秀而古”，“有真为假，做假成真”。《园冶》中将假山分为园山、厅山、楼山、阁山、书房山和池山等。为传承中国园林假山、叠山艺术，流芳园中布置了一座大假山，其特征为“池山”。

Long in Yu Garden in Shanghai. Exquisite stones or rocks enrich the scenery and even make the garden famous.

The appreciation of stones began with the appreciation of mountains in China. An ancient book of *Shan Hai Jing* or the *Classicsof Mountains and Seas* of Qin dynasty (221—206 BC) contains five chapters on mountains in all over of China. Confucius said: “The wise delights in waters, the benevolent delights in mountains.” In ancient times, emperors went up mountains and worshiped mountains; the imperial gardens in Qin and Han dynasties contained “the three mountains or islands” for inviting immortals; the scholars of Wei and Jin dynasties secluded in mountains; the scholars of Tang, Song and Yuan dynasties wrote landscape poetry and painted landscape paintings; in Ming and Qing dynasties, literary landscape gardens became popular. The sentiments of mountains have permeated the daily lives and emotions of the Chinese people. Mountains have become metaphors of personal traits and pursuits. Dai Xi (AD 1801—1860), a Qing dynasty painter, said: “Spring mountains are desirable for touring, summer mountains for looking at, autumn mountains for climbing, and winter mountains for residing.”

In traditional Chinese gardens, mountain and water are the skeleton of a garden. If there is no water, the garden is not complete; if no mountain or hillock, the garden would not be a garden. The work of piling up soil and rocks was called making *jia shan* or artificial rockery. It was a very important part of the Chinese garden. The art of making rockery has a long history of over two

中国园林中的假山，贵在“做假成真”之巧，妙在“似与不似”之间，成在“山色有无中”之境。园林假山既有自然真山之神采，又有造园者心目中之丘壑。中国叠山艺术深受中国山水画理论和技法的影响，五代画家荆浩在《笔法记》中提出“气、韵、思、景、笔、墨”的绘景之六要；宋代画家郭熙在《林泉高致》中提出山有“高远、深远、平远”之构图法；宋代画家马远和夏圭的“马一角”、“夏半边”的构图法，以及中国画中的画山石质所采用的诸多“皴法”，这些都是叠山所借鉴的方法。中国园林的叠石名家也多为画家，如明代张南垣、计成，清代戈裕良和石涛等都擅长山水画。中国现存最著名和最优秀的园林假山有两处，一是苏州环秀山庄内的太湖石假山，此山为叠山大师戈裕良的代表作品；二是扬州个园里的黄石假山“秋山”，相传此山出自石涛之手笔。前者，山势层峦叠嶂，曲水缭绕，蜿蜒迂回，石质圆润清秀，如画家郭熙的“卷云皴”效果。后者，山体峻峭凌云，高耸奇险，壮丽雄伟，石形棱角分明，如画家马远和夏圭的“大斧劈皴”的效果。两座假山尽得造化之妙，实为“虽由人作，宛自天开”之杰作！

在流芳园的总体布局中，临中心湖畔布置有一座大型假山，依土坡用太湖石叠石而成，以岩、峦、洞、穴与涧、壑、坡、矶相结合，形成一处层峦叠嶂、曲水缭绕、蜿蜒迂回、壮丽雄伟的“秋山”景致，与场地原有的山坡以及远处的群山浑然一体。沿着水际布置有

thousand years. In *Shang Shu* or *Book of Documents*, it recorded that “A lack of one basketful of earth spoils the entire effort to build a nine-ren tall mountain”. According to *Shi Ji* or *Records of Histories*, the imperial garden of Tai Ye Chi in Jian Zhang Gong of Han dynasty contained the three man-made immortal mountains of Fang Zhang, Peng Lai and Ying Zhou. In Song dynasty, emperor Huizong ordered the construction of the largest rockery garden (about 120 acres) during that time, Gen Yue, and its large man-made mountains built with earth and rocks. During Ming and Qing dynasties, the rockeries in the gardens of Jiangnan region were constructed mainly with Taihu rocks and yellow rocks. Rockery became the symbolic scenery of a garden at that time. Ji Cheng said: “Arrange rockery beside a pond to create attractive scenery of a garden.” In early times, man-made mountains in Chinese gardens were constructed with earth, and later half earth and half rocks, then mostly with rocks. In *Yuan Ye*, Ji Cheng summarized the different types of rockery, such as *yuan shan* in scholar gardens, *ting shan* in front of halls, *lou shan* next to storied buildings, *ge shan* in front of tower, *shu fang* shan next to studios, and *chi shan* by the ponds, etc. In order to inherit the tradition of man-made rockery in Chinese gardens, a large rockery mountain is laid out in Liu Fang Yuan, which is categorized into *chi shan*.

The best rockery mountains in Chinese gardens are the ones which are so artful that they look like real mountains, rather than exact imitations of the natural mountains. They should present the essence of real mountains and the conceptualized mountains of the designer. The art and craft of

扬州片石山房
The Rockery Villa in Yangzhou

湖石驳岸，蜿蜒曲折、错落有致、时隐时现，与波光、柳枝、倒影相映成趣，好似一幅山水画长卷。在几处庭院当中，布置有太湖峰石。沿着游廊和粉墙，布置有奇峰异石，形成丰富的“邻借”和“对借”的景观。在整个园林里面，假山和叠石与水系、花木、建筑融为一体，处处生景，时时变幻，诗情画意尽在其中。

rockery making was affected by the theories and techniques of Chinese landscape painting. For instance, Jing Hao (AD ?—936?) a Five Dynasties painter, summarized the six requirements for landscape painting in his book of *Bi Fa Ji or Notes* on *Brushwork Techniques*: *qi* or vital energy, *yun* or charm, *si* or composition, *jing* or subject, *bi* or brushwork, and *mo* or application of ink. Guo Xi of Song dynasty proposed three composition techniques: loftiness, depth, and horizontality. Most rockery designers in the past were painters such as Zhang Nanyuan (AD 1587—1671) and Ji Cheng of Ming dynasty, Ge Yuliang and Shi Tao of Qing dynasty. Two of the best rockery mountains existing in Chinese gardens today are the Taihu rock mountain in Huan Xiu Shan Zhuang in Suzhou and the yellow rock mountain in Ge Yuan in Yangzhou. The former was designed by Ge Yuliang, the later attributed to Shi Tao. They are the masterpieces of rockery hillocks in Chinese gardens, which presented the notion of "though man-made, it should look like something naturally created"!

In the overall layout of Liu Fang Yuan, a large rockery mountain is placed next to the central lake. It is situated at the foot of a hill on the western part of the garden. It is constructed with Taihu rocks, and considered the "Autumn Mountain" in the design of the Autumn Garden. The scenery of the mountain is composed of various rockery forms including cliffs, caves, gullies and ditches, etc. Inside the other courtyards, large freestanding peak Taihu rocks are placed as the main features, while the small ones are placed along covered walkways against the white washed garden walls. Part of the bank of the lake is constructed with Taihu rocks as well. Inside the garden, rockery and peak rocks are integrated with the buildings, water scenery and planting to present poetic and picturesque conceptions.

扬州个园 – "秋山"
Ge Yuan in Yangzhou - The Autumn Mountain

流芳园 – 叠水假山
LFY - Rockery and cascade

扬州瘦西湖 － 紫玉兰
The Slender West Lake in Yangzhou - The purple magnolia

嘉树繁花

中国幅员辽阔，植物繁多，品种各异。中国上古时期的园林是以“囿”和“圃”的形式出现，即以动物和植物作为园林的主要内容，如豨韦的“囿”，黄帝的“圃”，先秦时期的郑国原囿，秦国具圃。《周礼》曰：“九职二曰园圃，毓草木。”之后有“苑”和“园”的形式，内容包括山水、建筑和花木等。中国山水画论中不乏论述树木花草在绘画中的重要性，如唐代王维《山水诀》和《山水论》曰：“平地楼台，偏宜高柳映人家；名山寺观，雅称奇杉衬楼阁。”宋代郭熙《林泉高致》曰：“山得水而活，得草木而华。”明代计成《园

Chinese Garden Plants

China has a vast territory and a profuse variety of plants. In ancient China, the earliest gardens were called *you* and *pu* or plots of land for animals and plants. Later, landscapes and buildings were added to the gardens, which were then called *yuan*. Plants were important elements in Chinese landscape paintings as well. Wang Wei , a Tang dynasty painter and poet, wrote: “Tower and gazebo sitting on a flat ground will be suitable to be surrounded partially with tall willow trees; temples sitting on famous mountains will be elegant with exquisite cedar trees around.” Guo Xi of Song dynasty said: “Mountains will be vivid with waters, and splendid with vegetations.” The Chinese garden was first formally called *yuan lin* by Ji Cheng in his book of *Yuan Ye*. The word “yuan” means a fenced area with a building and a pond, and “lin” means forest or woods. So in Chinese,

冶》一书的开篇《兴造论》云："园林巧于因借，精在体宜。"这是第一次正式用"园林"统称造园，而其中的"林"则代表树木花草。由此可以看出，中国园林始终离不开树木花草。可以说，无花不成园，无木不成林。"园"与"林"结合为一体才能构成一座完整的园林。

在中国园林特别是私家文人山水园当中，植树栽花除了绿化环境的作用外，还具有三方面的审美功能：一是造景入画，二是时令变化，三是比德寓意。造景入画，须注重树木的姿态，树石的搭配，树木的位置和生长的特性，使其俯仰生姿，各具意态。陈从周先生在《苏州园林》中说："对于培花植木，必须深究地位之阴阳，土地之高卑，树木发育之迟缓，耐寒抗旱之性能，姿态之古拙与华滋，更重要的为布置的地位与树石的安排了。"

中国园林特别强调"象外之情、景外之意"，其中包括天时、季相变化之美。时令变化，须考虑树叶色彩的季节变化，开花季节的先后，气象变化与植物的形态关系。郭熙《林泉高致》曰："真山水之烟岚，四时不同，春山淡冶而如笑，夏山苍翠而如清，秋山明净而如妆，冬山惨淡而如睡。……春山烟云连绵，人欣欣；夏山嘉木繁阴，人坦坦；秋山明净摇落，人肃肃；冬山昏霾翳塞，人寂寂。看此画令人生此意，如真在此山中，此画之景外意也。"清代诗人汤贻汾《画筌析览·论时景》曰："景则由时而现，时则因景可知。"这里的"时"包括季节、时辰和天象，由"时"的变化而产生不同的景致，而不同的景致又反映出"时"

yuan lin or garden means an enclosed area of land with buildings, pond and plants inside. The full embodiment of a garden comes with plants.

Inside Chinese gardens, especially in private scholar landscape gardens, trees and flowers have three aesthetic significances: the picturesque effects, the seasonal changes and the symbolic associations to virtues. In the making of picturesque scenery, the garden designer looks for aesthetic shapes or postures of the plants, and composes them with other elements in the garden such as rocks and buildings. Chen Congzhou said in *Suzhou Gardens*: "In selecting and cultivating flowers and trees, meticulous care should be taken of their location, sunny or shady, their positions, high or low, their rate of growth, quick or slow, their resistance to cold, strong or weak, and their shape, quaint or magnificent. More important is the arrangement of their location and composition with rockery."

The Chinese garden emphasizes the implications of sentiments beyond imagery and meanings beyond scenery. There are, for example, the beauty of time and seasonal changes in the garden, the changing colors of leaves, the timing of blossoming, the changing pattern of shadows from trees, etc. Guo Xi said in *Lin Quan Gao Zhi* or *High Interest in Forest and Spring*: "The mist of real mountain and water is different in four seasons. The spring mountains are so light and elegant as a smile, summer mountains so greenish as if they were dripping water, autumn mountains so clear as with a makeup, and winter mountains so gloomy as if they were sleeping…… The unbroken mist in spring

扬州小番谷
Xiao Pan Gu in Yangzhou

的不同。

比德寓意，借物励志，是中国古代哲人将自然美与人伦道德相互对应、相互融合的一种文化现象，其侧重在中国文化所赋予植物的品格内涵上。以花木来比德，最著名的是孔子所说的："岁寒，然后知松柏之后凋也。"荀子曰："芷兰生于深林，非以无人而不芳。"经过历代文人墨客的提炼和文化沉淀，中国文人将多种花木赋予了拟人的品格，如"岁寒三友"，宋代林景熙《王云梅舍记》曰："即其居累土为山，种梅百本，与乔松、修篁为岁寒友。"宋代诗人曾端伯提出了"花中十友"：兰为芳友、梅为清友、腊梅奇友、瑞香殊友、莲为净友、栀子禅友、菊为佳友、桂为仙友、海棠名友、荼蘼韵友。又有清人张潮《幽梦影》中提出的：梅令人高，兰令人幽，菊令人野，莲令人淡，竹令人韵，松令人逸。最著名的还有以梅、兰、竹、菊代表"四君子"之寓意：梅，探波傲雪，高洁志士；兰，深谷幽香，世上贤达；竹，清雅澹泊，谦谦君子；菊，凌霜飘逸，世外隐士。另外，莲花，中通外直，不蔓不枝，"出淤泥而不染，濯清涟而不妖"，寓意品格高尚。这些花木不仅在园林中起到美化园林、

流芳园 - 翠绿出墙
LFY - The green coming out of the garden wall

mountain makes one feel happy; the trees and shades of summer mountain makes one feel candid; the bright and pure tinge of autumn mountain makes one feel solemn; and the haze and shade of winter mountain makes one feel lonesome. Looking at this kind of painting, one will have those feelings as if one is in the real mountains. These artistic conceptions of the painting are beyond the imagery on the painting." Tang Yifen (AD 1778—1853), a Qing dynasty poet, wrote in *Hua Quan Xi Lan*: "Scenery is presented over time, while time is known through scenery." The notion of time here includes seasons, time of a day and celestial phenomena. The change of time will make different scenery, while different scenery will present the change of time.

The symbolic and metaphoric implications of plants are a cultural phenomenon in traditional Chinese culture. The ancient scholars of China made parallel between human virtues and the things of natural beauty. Trees and flowers, for example, are often used to imply the morality and virtues of a person. Confucius said: "In cold days or winter, one will know that pine and cypress will wither after all other plants." Xun Zi (313—238 BC), a thinker of the Warring States period, said: "Orchids grow in deep woods; their fragrances are still there even if people don't notice them or not. Throughout the history, the Chinese scholars had selected and refined

丰富景观的作用，而且反映出中国美学思想注重“品”和“雅”的特性，体现了中国人文的内涵和精神境界。

中国园林有以花木之名作为园名的，其寓意也是突出植物的人文品格，以显示出园主人的思想境界。清代江南名园扬州“个园”就是以竹的特征来命名的。在清刘凤诰《个园记》中这样说：“主人性爱竹，盖以竹本固，君子见其本，则思树德之先沃其根；竹心虚，君子观其心，则思应用之务宏其量；至夫体直而节贞，则立身砥行之攸系者实大且远；岂独冬青夏彩，玉润碧鲜，著斯州筱荡之美云尔哉？主人爰称曰：‘个园’。”由此可见，中国文人喜好将树木花草赋予人文气息和文化内涵。在流芳园的设计当中，植物的选择和种植体现了这种文化传统。其中，一个设计重点就是将场地中现有的当地植物如加州橡树结合到园林布置当中，使其构成中国式的园林景观。另一方面，是使新种植的中国品种花木适应园中的场地和当地的气候环境，同时，在植物选择上还需考虑树木的姿态，使其能够“入画”而构成有画意的景致。

a variety of flowers and trees to be the metaphor for human morality and virtues. Lin Jingxi (AD 1242—1310), a Song dynasty scholar, said in the *Record of Wang Yun's Plum Cottage*: "In the dwelling where an earth hill is piled up, a hundred plum trees are planted together with pine trees and bamboos; all these become the friends of the winter." Zeng Duanbo, a Song dynasty poet, selected ten kinds of flowers to make metaphors for human friends: orchid as fragrant friend, plum pure, wintersweet peculiar, winter dsphne special, lotus as clean, gardenia the Zen, chrysanthemum excellent, osmanthus the immortal, malus spectabilis famous, and rosa rubus charm. The most popular metaphor is so called the "four gentlemen": plum, orchid, bamboo and chrysanthemum. Plum represents the strength and nobility, orchid humbleness and loftiness, bamboo simplicity and modesty, and chrysanthemum elegancy and seclusion. In Chinese gardens, therefore, the flowers and trees are not only for adding greenness and beauty, but also metaphors for positive and honorable human attributes. The aesthetic ideology of Chinese garden culture emphasizes the *pin* or quality and *ya* or elegancy of the scenery.

To name a garden after a plant is common in China. The purpose is to present the personal character of the owner or master of the garden. Ge Yuan in Yangzhou, for instance, was named after bamboo. Liu Fenggao (AD ?—1830), a Qing dynasty scholar, said in the *Record of Ge Yuan*: "The owner of the garden likes bamboo because bamboo has strong roots, so whenever seeing the bamboo, the gentleman will think that to establish moralities begins from nurturing the roots; bamboo is hollow inside, so when the gentleman looks at it, he will be reminded not to do things beyond his capacity; the body of bamboo is straight with joints, so the conduct of a person should be firm and far-sighted; bamboo is green in winter and variegated in summer like embellished jade,

梅 清 汪士慎 上海博物馆藏
Wang Shishen (1686-1759),
Plum, Shanghai Museum

兰 清 蒋廷锡 南京博物院藏
Jiang Tingxi (1669-1732),
Orchild,Nanjing Museum

竹 清 郑板桥 旅顺博物馆藏
Zheng Bangqiao (1693-1765),
Bamboo,Lu Shun Museum

菊 明 李因 旅顺博物馆藏
Li Ying (1610-1685),
Chrysanthemum,Lu Shun Museum

aren't they beautiful? Therefore, the owner named the garden as Ge Yuan or a Bamboo Garden." This shows that the Chinese scholars are fond of the cultural connotations and personifications of plants. The design of Liu Fang Yuan has continued this cultural tradition. Those implications have been applied to the selection and planting of flowers and trees inside the garden. Another important aspect in the design process is to integrate the existing plants such as the California oak tree on the site with the aesthetic composition of the scenery to provide the garden with an authentic Chinese style. At the same time, it's necessary to adapt the newly planted Chinese plants to the site of the garden.

总体布局

THE OVERALL LAYOUT OF THE GARDEN

中国园林历史悠久，内涵丰富，独具特色。造园思想一脉相承，而园林形式各异，其遵循的是“有法而无式”、“构园无格，借景有因”、“因地制宜”和“同中求异”的造园基本法则。陈从周先生提出了造园八字原则：“园以景胜，景以园异。”流芳园占地面积约为 75 亩，与现存中国文人山水园林的规模相比，是比较大的。如苏州最大的古典园林拙政园占地也只有约 72 亩，留园占地约 50 亩。在这样一个较大的空间范围内，如何布置建筑空间、庭院空间和山水空间，才能实现计成所说的“园林巧于因借，精在体宜”，这是一项颇具挑战的设计任务。陈从周先生提出过“小园以静观为主，动观为辅。大园以动观为主，静观为辅”的造园思想。流芳园占地较大，适合以动观为主。具体采用“园中有园”的设计手法，将各具特色的景点组织成片区，使大园不觉其空旷，从而达到曲径通幽、

The Chinese garden has a long history and a variety of distinguished features. The gardens were created with the same design principles, but distinguished from one another. There are rules but no fixed formulas on garden design and construction; there are no standards for designing gardens but certain principles of making use of the natural scenery, fitting in with the site and seeking diversity from the similarity. All these principles are the fundamental guidelines in the Chinese garden design. Chen Congzhou said: “A garden will be superior with its sceneries and views; the sceneries and views should be distinct in different gardens.” In Liu Fang Yuan, the total land area is about 75 mu or 12 acres. Compared with the existing scholar gardens in China particularly in Suzhou, Liu Fang Yuan is larger than most of them. The Humble Administrator’s Garden in Suzhou, for instance, has a land area of 72 mu, while the Lingering Garden only 50 mu. It is, therefore, quite a challenge to design Liu Fang Yuan within such a large area, for example, how to organize the architectural spaces, the courtyards as well as the landscape features, so as to realize the effect as Ji Cheng said in *Yuan Ye* that “a garden’s setting should ingeniously follow and borrow from the existing scenery and lie of the site, and be refined to fit the scale of the site and suitable for each other”. Chen Congzhou said: “In-position-viewing” should be predominant and “strolling-viewing” supplementary, and vice versa.” Liu Fang Yuan is suitable for strolling-viewing as predominant and in-position-viewing as supplementary. In addition, it is also suitable for organizing the garden elements and features into various clusters to create the feeling of a garden

1. 主入口区

   Main Entrance
2. 春园

   Spring Garden
3. 夏园

   Summer Garden
4. 峪园

   Ravine Garden
5. 宝塔园

   Pagoda Garden
6. 盆景园

   Penjing Garden
7. 秋园

   Autunm Garden
8. 冬园

   Winter Garden
9. 松涛园

   Pines Garden
10. 幽竹园

    Bamboo Garden
11. 汉庭湖

    Huntington Lake

流芳园总平面图

The illustraded Master Plan of Liu Fang Yuan

步移景异、景外有景的艺术效果。流芳园在总体设计上，根据场地的特质和形态，构思和布置出了九个景观区，取名为“九园”，包括“春、夏、秋、冬”四园，以及峪园、宝塔园、盆景园、幽竹园和松涛园。

中国传统造园，又称“构园”，“必先相地立基”，如同绘画中的“经营位置”，亦即现代人所说的总体布局。流芳园场地位于汉庭顿的中西部，根据“相地”的结果，整个园林以一个湖区为中心，其他各个庭院环绕其布置。流芳园中的主体厅堂建筑群和庭院，布置在湖之北面，即整个园之中心部位，形成“背山面水”、“坐北朝南”的态势。这块场地地势平坦，空间大小适合一个“二进庭院”。主体建筑群以“二进院落”围合而成，此组庭院建筑取名为“冬园”，其建筑主要包括：南面临水的“平湖秋月轩”和“同庆堂”[①]，中间为“江山多娇”戏楼，北面为“文澜鹂音阁”以及东西厢房组成。在“二进院落”中轴的东面，有一溪水流过，构筑一处“曲水流觞”的文人雅聚庭院空间，类似于晋代书圣王羲之的“兰亭序”中所描述的雅聚活动场所。庭院内以“流觞曲水”为主题景观，种植兰草，配以“兰亭”。在此庭院北部，将一棵巨大的橡树保留下来，以其为中心，围合成另一个庭院，取名为“古木风流”。

within a garden. The overall design of the garden should achieve the effects of a large space but not too open, strolling with changing scenes, and scenes beyond scenes. Based on the characteristics of the site, the master plan of Liu Fang Yuan is laid out with nine distinguished subordinate garden areas. In the early phase of the design, I gave the subordinate gardens thematic names. The “Nine Gardens” include Spring Garden, Summer Garden, Autumn Garden, Winter Garden, Ravine Garden, Pagoda Garden, Penjing Garden, Bamboo Garden and Pines Garden.

Gou yuan is another term for *zao yuan* or making a garden in the design of traditional Chinese gardens. It begins with the site survey and investigation and then situating foundations. It is like the arrangement of objects in Chinese paintings, which in modern parlance is similar to the master planning or layout. The site for Liu Fang Yuan is located on the western part of the Huntington ground. In the design, based on the survey and analysis of the site, the layout of the entire garden is centered with a lake and the buildings and courtyards surrounding it. The main buildings and courtyards of the garden are placed on the northern side of the lake so as to follow the Chinese tradition of “fronting water with mountains on the back” and “sitting on the north while facing the south”. The spot on the northern side of the lake is a flat area that is good for a two-courtyard layout. This complex is named the Winter Garden, where the main buildings include a waterside pavilion, an opera tower, and a two-story library building. On the eastern side of the main courtyard, there is

①景点的名称为设计过程中暂时之用。

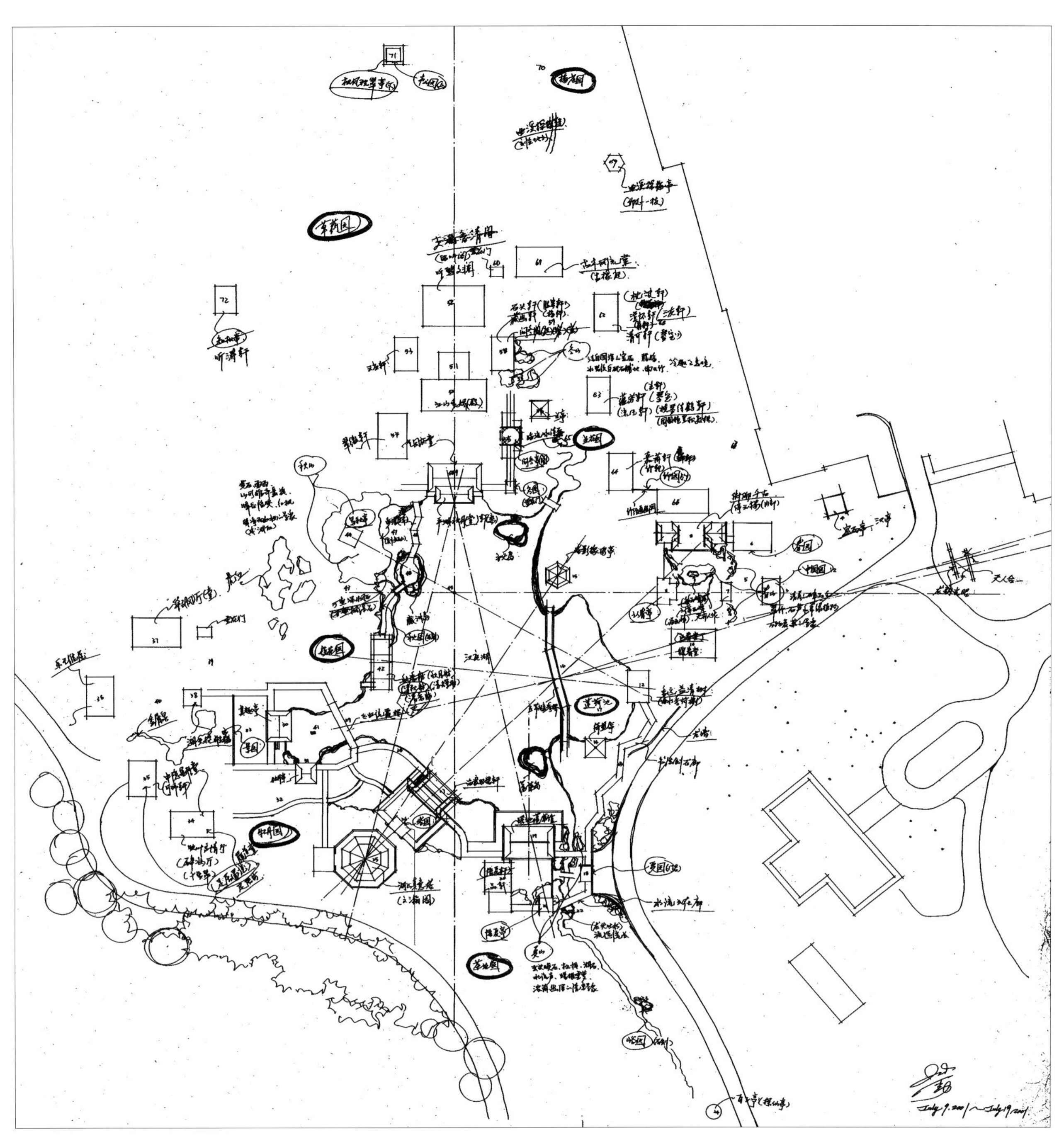

流芳园景观点布置草图

The sketch plan of scenery in Liu Fang Yuan

a natural stream running through this area from north to south. The design utilized it to make a very famous Chinese scholar gathering place *qu shui liu shang* or “winding stream with flowing wine cups”, which originally came from the legendary event held by Wang Xizhi, a calligraphy master of Jin dynasty. This scene is planned as a retrospect of the famous gathering of scholars at the Orchid Pavilion in Shaoxing in A.D. 353. On the north of this spot, there is a huge California oak tree to be saved to make a distinct garden feature. A courtyard is planned to surround the tree. This scene is named “the distinguished and admirable old tree”.

流芳园 – 门楼砖雕
LFY - The brick carving of the Garden Name - "Liu Fang Yuan"

在中心湖区的东面布置了另一组庭院建筑，取名为“春园”。这一处与汉庭顿的主入口区相连，是进入流芳园的主入口庭院。“春园”是一个东西向“二进院落”，规模比北区的主体院落略小。东面布置一座入口庭院，由入口建筑“明慧轩”和一个砖雕门楼围合而成。中心庭院以一座巨大的太湖石峰为主景，石名取为“停云峰”，孤峰突起，独立寒天，从容自若，寓意顶天立地之人格风范。石峰南边设置一个小水池，石峰倒映其中，更显挺拔俊秀。中心庭院四周布置有“探春亭”、“锦云堂”和“俯仰千古楼”。整个春园的峰石碧潭和庭院建筑构成一处“峰庭虚明”的景致。穿过“春园”，沿着游廊向西进入中心湖区，空间顿觉开朗与壮观，形成一组“欲扬先抑”、“小中见大”和“柳暗花明”的空间艺术效果。

On the eastern side of the central lake, a series of buildings and courtyards are planned for the Spring Garden. This area is close to the Huntington's main entrance. It is planned to be the main entry gate to the entire Liu Fang Yuan. The Spring Garden is a double-courtyard with an east-west orientation. It is smaller than the Winter Garden complex. The first courtyard is enclosed by an entry gate building which is named the "Knowing Wisdom Chamber", a brick carved gateway, and white-washed walls. On its western side, there will be a giant free-standing Taihu rock sitting in the main courtyard as the main scenery named the "Peak Rock Stopping Clouds". This peak rock symbolizes the upright merit of human being with its stance of independence and composure. A small pond sits in front of and reflects the rock. The main courtyard is surrounded with the "Exploring Spring Pavilion", the "Bright Clouds Hall" and the "Seeing-through-ages Tower". This complex of the peak rock and deep pond as well as the buildings and courtyards make the vivid scenery of "peak rock court unveils real and imaginary scenes". Walking through the Spring Garden to the west, passing a grove of oak trees, one will see the central lake and suddenly feels the spaciousness and

“春园”可有“小园几许，收尽春光”（秦观《行香子》）之气韵。

从“春园”向南，通过一个曲廊，下到“曲水香荷”水榭一处（“爱莲榭”系建成后的名称），再向西望去，一座玉带石桥横贯湖上，三孔石桥将人的视线连接到湖的中央水面。玉带桥的两端各设有一座园林建筑：左边是一个“绮望亭”（“玉镜台”）方亭，右边是一个六角形的两重檐阁楼（“三友阁”）。四座建筑（亭、榭、桥、阁）围合成一个湖中湖，水中植莲荷，夏季荷香飘溢，构成一处别致的景区。继续向南，进入“夏园”，一个“一进半”的庭院，并设有入园的次入口（现为入园的主入口）。庭院北面设一个四面厅堂（“玉茗堂”）和一个亲水平台，此平台面向流芳园北边的圣盖博群山，故取名为“邀山台”。这里是品茶、观湖、观山的好景点。庭院南面设有斋室一个（“清芬斋”），飞梁一座，横跨于溪流和瀑布之上。沿溪流继续南下设茅亭一座（“涤滤亭”），回首仰望，好似云水相接，飞天直下，实有“水流云在”之意境。

位于“夏园”西边的，是一个较高的山坡，这里是流芳园中的至高点。在这里布置有一座七层重檐的宝塔，其用意有三：一是创造一处登高望远的佳处，可眺望北面巍巍群山，可观赏全园湖光山色以体会“湖山真意”的意境；二是创造一个全园的“点景”建筑，亦成为一个“借景”的建筑；三是将“宝塔”这一中国风景区常见的建筑置于中国园中，展

openness. The design makes the effects of seeing the large from the small as well as the shades of willows and bright flowers. The Spring Garden, though small, captures all the charm of the spring.

Strolling to the south, passing through a winding covered walkway, the visitor arrives at a waterside pavilion named “Love for the Lotus Pavilion”. Looking towards west, there is a three-arched stone bridge lying on the water of the lake. On the southern end of the bridge, there is a square shaped pavilion named “Terrace of the Jade Mirror”, and a hexagon “Pavilion of the Three Friends” on the northern end. These four garden structures make an enclosure of the water space, which becomes a small lake within a lake. Continuing to the south, one arrives at the Summer Garden, a one-and-half courtyard complex with the second entry gate (the current main entrance) to the entire garden. The major building is a four-sided hall now functioning as a tea house named “Hall of the Jade Camellia”, located on the northern side of the courtyard with a waterside platform on its north. This platform faces the lake and the northern distant San Gabriel Mountains, so it was named the “Terrace that Invites the Mountains”. This spot is a good place for enjoying tea and viewing the lake and mountains. On the southern side of the courtyard, a studio and a covered walkway sit on top of a cascade stream and rockery. The stream tumbles down to the south into a gully. A thatched small pavilion named “Pavilion for Washing Away Worries”, is placed on the bank of the stream. From this spot looking back to the cascade, one can find the water fall and the clouds in the sky seem to be merged into one and poetic with the “tumbling waters merge into the upper clouds”.

流芳园总体鸟瞰草图
The sketch aerial view the whole garden of Liu Fang Yuan

示出中国传统建筑的独特艺术魅力。

在宝塔的东北面设一处庭院，沿湖布置有环廊，信步其中，顿生“沧浪环碧”之意境。由“塔园”的环廊向西北，进入一个“水庭”。一座木廊虹桥跨过水上，好似一道“彩虹”浮于湖上。水庭西面设有一个高台轩馆（“天趣轩”），南面设有“牡丹亭”一座，周边山坡上遍植牡丹。牡丹亭内可作为小规模音乐表演场所，借助水庭中的水面回音效果，创

There is a high ground on the western side of the Summer Garden. It is the highest spot in Liu Fang Yuan. In the master plan, a seven-story pagoda or tower sits on top of the hill. There are three intentions in the design of the pagoda: first, to create a higher view point best for looking afar towards the high mountains in the north and for viewing the entire Liu Fang Yuan to capture the true meaning amidst the lake and mountain; secondly, to create a point-of-view structure and a scene for borrowing from within the garden; and thirdly, to represent the exquisite form of pagoda, which is one of the unique Chinese traditional architectural features that commonly seen in Chinese landscape scenic areas and large gardens.

造一处观戏听曲的佳处，以再现著名昆曲《牡丹亭》的韵味。至此，整个南面环湖景区恰好形成了北部主体庭院向南观望的“对景”——湖光塔影。

“水庭”西面，沿缓坡设置一个盆景园，以曲折回廊连接各个展厅和景墙，围合而成的庭院内布设有山石、涌泉、花木和嘉树，营造出“壶天揽胜”的意境。由“水庭”向北，就进入了流芳园中最大的假山群，即秋园。大假山以中国园林著名的太湖石为主要材料构筑而成，并布置一叠水瀑布沿假山而下，呈现出丘壑深峪、万峰叠嶂、飞瀑连天的山水画意。假山上设有“秋月亭”一座，周边种植枫树、桂花等秋景树林。在假山山麓临湖处布置一座石画舫、一个“瀛洲岛”、一座“鱼乐”曲桥和一个“印月”石塔于湖上，以此构成一处观石、观叶、观鱼和观月的独特景区。

在流芳园的最北面，是保留下来的一片松树林。在林间布置小径若干，“听涛轩”一座，“松风积翠亭”一座。此处是全园最疏朗的景区。在其东面是一片竹林，翠竹茂密，设曲径竹亭于林内，通幽之处，别有洞天。整个北部片区为全园创造了一个苍松翠竹的背景。

On the northeastern side of the pagoda, there is a courtyard enclosed with curved covered walkways which sit along the bank of the lake. Further to the northwest, a Water Court is designed for music performance. It is composed of an arched wood bridge across an outlet of the lake, a gazebo sitting on a tall terrace, a pavilion by the water and the covered walkways. The arched wood bridge looks like a rainbow over the water. The gazebo is named "Pavilion of the Heavenly Interests". The pavilion is named “Peony Pavilion” which recollects a famous Yuan dynasty Kunqu Opera, *Peony Pavilion*. The design of the Water Court aimed to utilize the echo effect of water to create a unique stage setting for music performances in the garden. At this point, the layout of garden features around the central lake on the southern portion has created perfect focal scenery for viewing the scenes of the lake and pagoda from the main courtyards on the northern side of the lake.

On the west of the Water Court, the Penjing Garden is situated on the eastern slope of a hill facing the lake. It consists of exhibition halls, corridors and walls, rockery, spring and cascade, and trees and flowers. The Penjing Garden is a place of “seeing all the splendors in a gourd-like world”. On the north of the Water Court, a large rockery mountain and a few garden features make up of the Autumn Garden. The rockery mountain or hillock is constructed with Taihu rocks. A waterfall is cascading from the rockery down to the lake. A pavilion named “Autumn Moon Pavilion” sits on top of the hillock, where foliage trees such as maple trees and osmanthus trees are planted around it. At the base of the rockery mountain, a boat-shaped pavilion sits by the lake where an islet and two stone bridges connect with each other. One bridge is named “Bridge of the Joy of Fish” and the

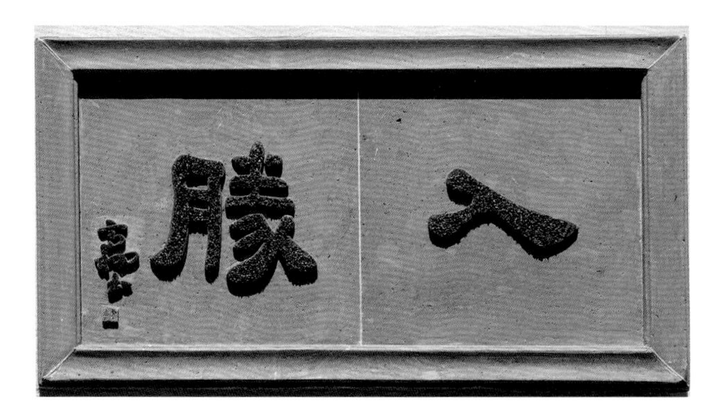

流芳园 － “入胜”砖雕
LFY - brick carving of "Ru sheng"

流芳园的总体布局按照中国传统园林法则营造，形成以湖沼为中心，将厅堂亭榭、叠山理水、曲廊横桥、秀木莳花的布置与场地的地貌、植被和水系紧密结合，创造出一个以小见大、园中有园、高低起伏、体宜因借的园林格局。这种独特的格局产生出别致的园林景观，从而达到“景以园异”和“因借无由，触情俱是”的效果，使得流芳园成为一座个性鲜明的传统式中国园林。

other the “Moon Bridge”. A stone lantern is in the lake next to the islet. All these make up a place best for watching the moon and carps inside the whole garden.

The most northern part of Liu Fang Yuan is a preserved pine forest. This area is designed as the Pines Garden with secluded paths winding among the trees, and a couple of pavilions are planned on the sides of the paths. One pavilion is named “Listen to the Waves of Pines”, and another the “Pines Breezes Gather Greens”. This Pine Garden is the most spacious garden inside Liu Fang Yuan. On the eastern side of the garden, there is an existing grove of bamboos. The design creates curved paths and a bamboo pavilion to make a tranquil place for resting in the deep greenery. The entire northern area becomes the green background of Liu Fang Yuan.

The master plan of Liu Fang Yuan was articulated in accordance with the design principles of traditional Chinese gardens. The garden is centered with a lake surrounded with buildings, courtyards, rockeries, bridges as well trees and flowers, etc. The overall spatial configuration is created with some of the typical concepts as in Chinese landscape gardens, among them are the concepts of seeing the large from the small, garden within gardens and borrowing from scenery. All these garden elements and features are integrated with the unique characteristics of the site so that the scenes are different from other gardens because the site and the sentiments generated from the site are different. Liu Fang Yuan is, therefore, distinctive and consistent with the traditional Chinese garden style.

流芳园 - 清越台
LFY - Qing Yue Tai Pavilion

九园十八景
THE NINE GARDENS AND EIGHTEEN VIEWS

流芳园的总体设计，在九个景观分区或“九园”布置的基础上，进一步创造了十八个核心景点，我将它们归纳成“九园十八景”。这十八景暂时取名为：峰庭虚明、塔影橡坞、曲水香荷、玉带接秀、万壑枫深、平湖秋月、文翰音清、古木风流、壶天揽胜、飞虹流霞、湖山真意、沧浪环碧、水流云在、幽谷清心、竹径通幽、曲溪探梅、松风涧水和积翠听涛。“九园”和“十八景”的题名均为初步的名称。在造园过程中或建成之后，这些题名将被进一步斟酌、完善和确定。

In the master plan of Liu Fang Yuan, eighteen distinct views or scenes have been created along with the subordinate “Nine Gardens”. I combined the gardens with the distinct views into an assemblage of “The Nine Gardens and Eighteen Views”. These so called “Eighteen Views” include the “peak rock court unveils real and imaginary scenes”, “reflection of pagoda in the oak dock”, “winding waters hold the fragrant scent of lotus”, “jade-ribbon-bridge linking beauties”, “myriad ravines in deep maples”, “autumn moon over the tranquil lake”, “the literary and musical realm”, “distinguished and admirable old tree”, “seeing splendid in a gourd-like world”, “flying rainbow and flowing red clouds”, “true significance of lake and mountain”, “azure water surrounded with bluish-green”, “tumbling waters merge into the upper clouds”, “secluded ravine and clear mind”, “path in bamboo grove leading to the hidden”, “searching out plum along a crooked stream”, “pine breeze and mountain stream”, and “listen to the waves of emerald pines”. All the names assigned here are temporary. They will be refined and selected during or after the construction of the scenery.

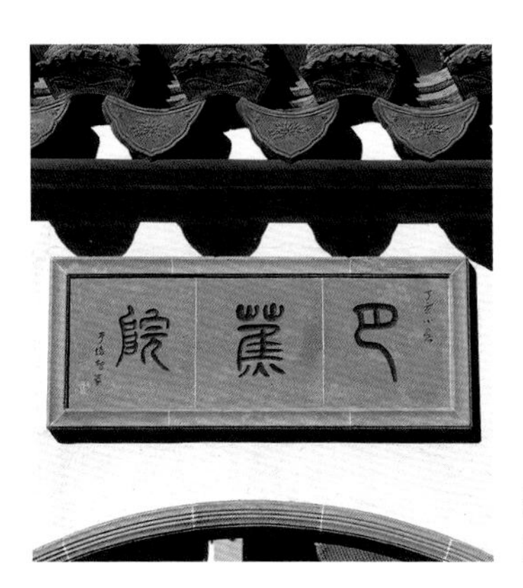

流芳园 - 芭蕉院门额
LFY - Name board of the Plantain Court

流芳园一期效果图

The Rendering of the Phase One of Liu Fang Yuan

春园：峰庭虚明、塔影橡坞

根据总体设计，春园位于整个流芳园的东部，是全园的主入口区。最东面的入口处设有一个石牌坊，与入口大门建筑围合成一个小的集散空间，在其北侧安置一座小亭，此亭取名为“汉亭”，亭内置一块“园记碑”，上面将刻有流芳园的造园历程、整个团队人员以及支持流芳园建设的捐款人名录等。这几个园林小品建筑与现场保留的几株参天雪松和柏树点缀了入口小环境，使游人在入园之前便可初步感受到中国园林的韵味。

入口大门是一座三开间的歇山顶建筑，进入后到达一个小院，院内一座三开间的硬山顶建筑作为“流芳园”的介绍厅。小院西侧设有一个砖砌的门楼，门额上方有“天开人作”的砖雕匾额。穿过门楼进入到春园的主庭院，以“先抑后扬”的手法，逐步展开。主庭院南北朝向，北面设有一个二层楼阁，取名为“俯仰千古楼”，南面设一个厅堂，取名为“锦云堂”，中间为一个水池，取名为“停云池”，临池北面立一座高耸的太湖石峰，取名为“停云峰”，意指高耸入云或像一片停着的祥云，类似于苏州留园中的“冠云峰”。“停云峰”是春园里的主题景观。古人曰：“石为云根”。峰石在中国园林中常常起到“点景”作用，而一块名石更可成为一座园林的“镇园之宝”。计成曰：“瘦漏生奇，玲珑生巧”，“片

Spring Garden: peak rock court unveils real and imaginary scenes, reflection of pagoda in the oak dock

In the master plan, the Spring Garden is located on the eastern side of Liu Fang Yuan. It is the main entrance area to the entire garden. Several garden elements make up the entrance space, including a stone arch, a small pavilion named *han ting* or “Han Pavilion”, where a stone tablet inscribes the record of the garden history and the names of donors, an entrance hall to the garden, and a few big cedar trees to be preserved as part of the garden scenery.

The entrance hall is a building of three-spans with a pitched roof. After entering through the building, the visitor comes to the first courtyard and on the northern side, another three-span building with a pitched roof is the introduction hall to the garden. On the western side of the courtyard, there is a brick carved gate with an inscribed tablet of *tian kai ren zuo* or “man-made but look like naturally created”. Passing through this gate, the visitor arrives at the main courtyard of the Spring Garden. The courtyard is oriented in north-south direction. A two-story building named Seeing-Through-Ages sits on the north, with a one-story hall named *jin yun tang* or “Bright Clouds Hall” on the south. In the center, there are a small pond and a giant peak Taihu rock. The free-standing peak rock is named *ting yun feng* or “Peak Rock Stopping Clouds”, which means towering up to clouds or a floating auspicious clouds. The ancient Chinese said: “Rockery is the root of

春园平面图
Plan of Spring Garden

1. 石牌坊
Stone Arch
2. 汉亭
Han Pavilion
3. 主入口
Main Entrance
4. 展厅
Exhibition Hall
5. 门楼
Entry Gate
6. 停云峰
Ting Yun Peak Rock
7. 停云池
Ting Yun Pond
8. 俯仰千古楼
Viewing Pavilion
9. 探春亭
Tan Chun Pavilion
10. 锦云堂
Jin Yun Hall
11. 橡坞
Oak Grove

山多致，寸石生情”。陈从周先生也常说：“名景名石是园林的命根子。”按照设计构思，“停云峰”应具有“透、漏、透、皱”的赏石特征。朱良志先生说：“瘦在淡，漏在通，透在微妙玲珑，皱在生生节奏。”“停云峰”如行云流水，风骨刚健，嶙峋风姿。它将有“独拔群峰外，孤秀白云中”（高丽定法师《咏孤石》）的气势。在此庭院置一块高大而秀美的太湖石，具有“伟石迎人，别有一壶天地”之境界以及“潭影空人心”之意趣。同时，在游人入园之后顿感庭院的苍古韵味和园林的历史感，“奇峰括天下之美，藏古今之胜，于斯尽矣”（宋代张淏《艮岳记》）。中国艺术有“好古”之情趣。中国园林之妙在

clouds.” Peak rocks are placed in Chinese gardens as focal-views. An exquisite peak rock can become the most precious feature of a garden and make the garden famous. Chen Congzhou said: “Exquisite scenery and rocks are the lifeblood of gardens.” The “Ting Yun Feng” peak rock should have the four qualities of an exquisite rock according to the Chinese aesthetics, *shou*, *lou*, *tou* and *zhou*, or slenderness, pierced, openness and wrinkles. Zhu Liangzhi said: “Slenderness shows the lightness, leakage the penetrability, openness the subtlety, and wrinkles the rhythm.” When visitors come to this courtyard and see this exquisite peak rock, they will feel something ancient and historical. The Chinese people have sentiments and delights concerning old things. Old trees, quaint rockery, a lone pavilion and deep water pond give a sense of historical and enchanted. Next to *ting yun feng* on the western side, there is a small pavilion named “Exploring Spring Pavilion”. A

春园透视草图
The sketch perspective of the Spring Garden

于苍古，没有古相，便无生意。古木、苍石、孤亭、深潭皆显古韵。这里的“古”是恒久之意，生机永存之意。临“停云峰”西面设有一小亭，取名为“探春亭”。其周围种有梅花、玉兰等春花树木，在春季到来时有浓郁的春天气息，春花满树，倒影池中，水天白云，峰石矗立其间，顿觉空间阔展，气象万千，生机盎然。整个庭院景致清秀古雅，虚实相生，气韵生动，故取名为“峰庭虚明”。

few spring flower trees are planted around it to enhance the spring atmosphere when they blossom during spring season. The scenery in this courtyard is quaint and vivid, which is named “peak rock court unveils real and imaginary scenes”.

Passing the pavilion and walking in the covered walkways, then going through a moon gate and stepping down to a grove of oak trees, the visitor vaguely sees the lake. Going on to walk to the edge of the lake, one sees the open spaces of the lake and the oak trees and willows as well as the reflection of a distant pagoda in the water. This scenery is named the “reflection of pagoda in the oak dock”. The characters of the spatial organization in the Spring Garden are highlighted by the

春园“峰庭虚月”

Spring Garden - the peak rock court unveils real and imaginary scenes

春园“塔影橡坞”

Spring Garden - reflection of pagoda in the oak dock

临“探春亭”西面设有一小亭，取名为“探春亭”。其周围种植梅花、玉兰等春花树木，在春季到来时有浓郁的春天气息，春花满树，倒影池中，水天白云，峰石矗立其间，顿觉空间阔展，气象万千，生机盎然。沿“探春亭”和游廊穿过一个月洞门，下数级台阶后进入到由几株大橡树形成的小丛林，隐隐约约看到中心区的湖水。继续向西来到湖边，空间顿觉开朗，橡树、柳树枝条映于湖上，婉约多姿，水波荡漾，平冈曲坞。远处宝塔倒印在柳影、水波和山影之间。此处景致取名为“塔影橡坞”。春园的整个空间特色是小中见大、虚实相生，俯仰相应，变化丰富。以“峰庭虚明”和“塔影橡坞”构成春园的两个主题景观。

so called “seeing the large from the small” and “growing out of the real and imaginary”. There are also views above and below in the garden. The two major scenes make up the Spring Garden: “peak rock court unveils real and imaginary scenes” and “reflection of pagoda in the oak dock”.

流芳园 － 傍晚景色
LFY - An evening view of the garden

夏园：曲水香荷、玉带接秀

由“春园”西面的曲廊向西南步行下到坡底，来到湖边，便进入夏园景区。此处一片湖光山色，尽收眼底。临水设有一座水榭取名为“爱莲榭”，水中种植莲荷，夏季荷花绽放，飘香四溢，风月清新。此水榭是观赏荷花的佳处，小榭风卷绣帘重，故取名为“曲水香荷”。从水榭向西望去，一座石桥横卧于湖上，三个桥孔将桥两边的湖面联系在一起。平缓的桥面像一枚玉带横落水上，将桥两头的景致连接在一起，一边是一座重檐六角阁（“三友阁”），另一边是一座四孔月洞门的亭子，取名为“绮望亭”（“玉镜台”）。亭子，得“亭台突池沼而参差”之美，在中国园林当中是观景和休憩的地方，也有“点景”的作用。计成曰：“花间隐榭，水际安亭，斯园林而得致者。”明代沈周曰：“天地有亭，万古有此月。一月照天地，万物辉光发。不特为亭来，月亦无所私。” 亭子，在中国园林中乃至中国山水画里，都是饶有深意的点景元素。

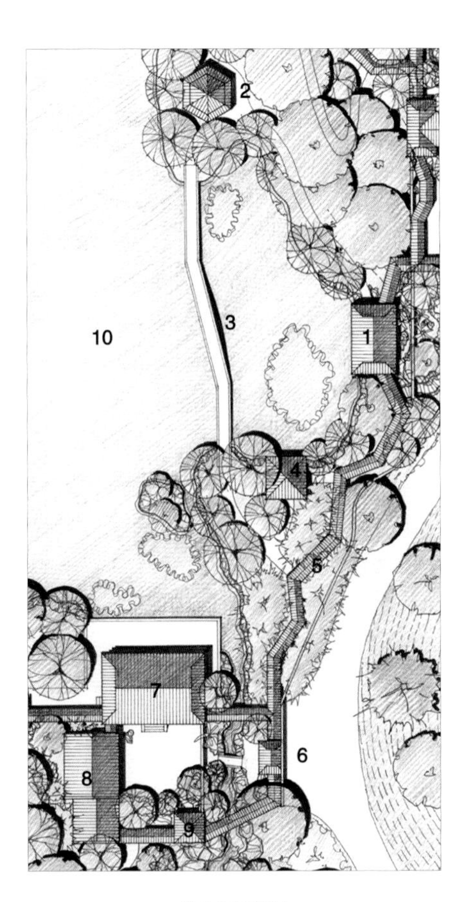

夏园平面图
Plan of Summer Garden

1. 曲水香荷榭（爱莲榭）
   Lotus Pavilion
2. 六角阁（三友阁）
   Hexagon Pavilion
3. 玉带桥
   "Jade ribbon" Bridge
4. 绮望亭（玉镜台）
   "Jade Mirror" Pavilion
5. 曲廊
   Zigzag Corridor
6. “流芳园”次入口
   Liu Fang Yuan entrance
7. 四面厅（玉茗堂）
   Four-sided Hall
8. 餐饮部
   Catering room
9. 精芳斋
   Jing Fang House
10. 汉庭湖
    Huntington Lake

Summer Garden: winding waters hold the fragrant scent of lotus, jade-ribbon-bridge connects beauties

Wandering along the corridor from the Spring Garden down to the Summer Garden area by the lake, the visitor sees a panoramic view of the lake and mountains. There is a waterside pavilion named “Love for the Lotus Pavilion”. Lotuses are planted in the water nearby. When the lotuses blossom in summer, fragrance overflow in the air, making the place the best spot to watch the lotus and enjoy the scent of the lotus fragrance. This scenery is named “winding waters hold the fragrant scent of lotus”. On the west, there is a three-arched stone bridge lying on the water of the lake. On each end of the bridge, there is one pavilion, the hexagon “Pavilion of the Three Friends”, and the square shaped pavilion named “Terrace of the Jade Mirror” distinguishable due to its four circular moon gates inside. Pavilion is one of the typical Chinese garden structures. It is for viewing scenery and resting, and serves as a point of view in the garden. Ji Cheng said: “A pavilion hidden in flowers or by the water, brings the garden delicacy”. Pavilion is also a common element in Chinese landscape paintings.

中国的石桥独具魅力，在宋代之后，园林中的桥梁多为石桥。夏园中有一座三孔石桥和一座一孔石桥，两桥由小岛连为一体。设计立意借鉴中国北方园林较为厚重的石桥样式，端庄典雅，浑朴清秀，与流芳园周围的大树和远山相映成趣。江南园林中的石桥多为单孔或石梁平桥。江南小镇上常有三孔石桥，但孔都较大，因为要通过船只。这座石桥设三孔有助于在视觉上将桥两边的水连成一体，有分有合，隔而不断。三孔和一孔桥也是中国石桥最多的样式，代表中国石桥的特征。石桥的桥孔与亭子的四个月洞门，构成一个“别有洞天”的意境空间。两座飞桥连接如带，后拥全湖；两岸柳木交映，水光日影，湖波澄碧，沁人心脾。此景故取名为“玉带接秀”。

Since the Song dynasty, the bridges in Chinese gardens have been mostly made of stone. In Liu Fang Yuan, the stone arched bridges are designed in accordance with the typical three-arched and one-arched bridges in China. In fact, the design of a few bridges in Liu Fang Yuan has referred to some in northern China, so that they are a little thick, dignified and austere, fitting the dry climate and the landscape of large trees and mountains in the Huntington area and beyond. The bridges around the central lake in Liu Fang Yuan are linked to exquisite pavilions, rockery and beautiful trees. These sceneries are named the “jade-ribbon-bridge connects beauties”.

夏园透视草图

The sketch perspective of the Summer Garden

夏园“玉带接秀”

Summer Garden - jade-ribbon-bridge linking beauties

夏园“曲水香荷”

Summer Garden - winding waters hold the fragrant scent of lotus

由“绮望亭”(“玉镜台”)向南沿东边的曲廊可至夏园的主体建筑群。曲廊隐在花木中，与其东面的“云墙”(“景云壁”)形成一个忽近忽远、忽明忽暗的空间变化，“随形而弯，依势而曲”，自有一番情趣。明代程羽《清闲供》曰：“门内有径，径欲曲。室旁有路，路欲分。”计成在《园冶·房廊基》中云：“任高低曲折，自然断续蜿蜒，园林中不可少斯一断境界。”夏园主体建筑是临水的一座四面厅，主景是向北望去的湖光山色，绿岛浮莲。北面远处的楼宇倒映在湖上，与楼宇背后的松林和远处的山脉交织成趣，幻影般的景致真有“蓬莱仙境”的气象，故此景取名为“瑶池蓬莱”。四面厅是一座歇山式园林建筑，四面设落地“合和窗”，以利于四面观景。其北侧设一平台挑在水面上，盛夏之季，荷花盛开，坐在旷然平台之上饮绿听香，周边景色尽收眼底，粲若绘境，摄魂荡魄，真有“纳千顷之汪洋，收四时之烂漫”的意境。

四面厅南边有一个庭院，庭院东面是一个入园的小院，庭院与小院通过一个月门洞相连，入口小院中有一溪流穿过。这个入口是整个中国园的次入口(目前作为主入口使用)，由一座云墙(“景云壁”)、八字形高墙、一座砖雕门洞和亭廊构成。门洞上有“流芳园”园名砖雕。粉墙黛瓦，一派江南园林气象。造景采用“抑景”手法，以景墙和曲廊分隔空间，形成“小中见大”、“别有洞天”的意趣。中国园林有“占尽风情向小园”之说。朱良志

Between the “Love for the Lotus Pavilion” and the main courtyard of the Summer Garden on the south, there is a zigzag covered walkway named *shui yun lang* or “Corridor of Water and Clouds”, which connects the two spots together. It is situated amidst the trees and flowers. A *yun qiang* or “cloud wall” sits next to the walkway. When strolling in the zigzag walkway, the visitor sees different scenes with the change of directions and spaces; every step brings a new view. This is a quite delightful experience. In the Summer Garden, the major building or “Hall of the Jade Camellia” has windowed walls on four sides, where the visitor can see the garden scenes on all four sides. A waterside platform on its north faces the northern distant San Gabriel Mountains, so the scenery is named “Terrace that Invites the Mountains”.

On the eastern side of the Summer Garden, there is a small entry courtyard through which a stream flows from north to south. The entry gate in the master plan is a secondary entrance, but now it is the main entrance to Liu Fang Yuan. This small courtyard is composed of the *yun qiang*, a tall entrance wall with a brick carved gate, a half-size pavilion, covered walkways, a white-washed curved wall, a stream and rockery. This area is designed with a typical Jiangnan or Suzhou garden style. It truly annotates the garden design concept of “seeing the large from the small”. Zhu

夏园“瑶池蓬莱”
Summer Garden - The Yao-chi wonderland

先生说：“园不在乎小，而能生烟万象，表现造化生机。”宋人冯多福曰：“以适意为悦。”夏园的整体空间主次分明，远近呼应。景致设计深奥曲折，通前达后，处处生出幻境。

Liangzhi said: “It doesn’t matter if the garden is small, as long as it creates a variety of imageries and the vitality.” “Delight in whatever suitable”, said Feng Duofu a poet of Song dynasty. The Summer Garden is designed with a variety of delightful scenes and distinguished garden spaces.

峪园：水流云在，幽谷清心。

流芳园中的湖水经由一条曲涧穿过夏园，再向南流下，经过一个石掇的悬壁和溪谷流向汉庭顿的日本园。悬壁落差约 7 米，溪谷长约百米，东西两侧特别是西侧，有大香樟树多株，下面以茶花树为主，并有梅花、樱花树穿插其中。设计中利用地形高差，构筑叠石水瀑，曲涧引岫，白水悬流，水声沸然。从谷底仰视，白水与白云融为一体，形成“水流云在”的景致和意境，取唐杜甫《江亭》中“水流心不竞，云在意俱迟”的诗意。沿着溪涧两岸，选秀石高下散布，不落常格，而依画理。坡上有山茶、梅花、香樟等多种花木，备四时之景色。此处翠盖丹英，错杂如织，潺潺流水，幽静清新。在此溪谷下端溪涧蜿蜒处，布置一座茅亭(“涤虑亭”)，坐在其中，使人忘却烦恼和忧虑，心肺畅然，实为一处“幽谷清心”的佳境。唐王维有诗曰:“行至水穷处，坐看云起时”。

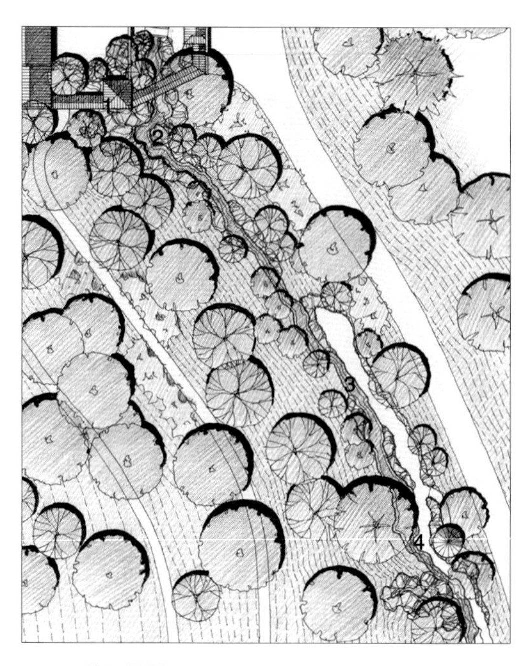

峪园平面图
Plan of Ravine Garden

1. 精芳斋
   Jing Fang Pavilion
2. 叠水（水流云在）
   Water Cascade
3. 溪谷
   Gully
4. 茅亭（涤虑亭）
   Thatched Pavilion

Ravine Garden: tumbling waters merge into the upper clouds, secluded ravine and clear minded

There is a stream running through the entire garden of Liu Fang Yuan from the north to the south and continue to the Japanese Garden further down to the south. The drop from the level of the lake to the very south end of Liu Fang Yuan is about 7 meters. The stream tumbles down into a gully or ravine, which is about 100 meters long extending till the Japanese Garden. On both sides of the stream in the gully, there are many camphor trees, camellias, cherry trees and plum trees. The design of this gully area is composed of a cliff with a cascade, the winding stream with rockery banks, and a thatched small pavilion. The cliff is constructed with Taihu rocks. This scenery is named as *shui liu yun zai* or “tumbling waters merge into the upper clouds”, extracted from Du Fu’s *Jiang Ting* poem: “The water is flowing, but my heart doesn’t compete with it; the cloud is floating, just like my mind.” The thatched pavilion is located by the stream on the southern end of Liu Fang Yuan. The lush plants along the stream seclude and freshen up this area. This lovely pavilion provides a spot for refreshing the visitor’s mind. The pavilion is named “Pavilion for Washing Away

峪园“水流云在”
Ravine Garden - tumbling waters merge into the upper clouds

陶渊明《归去来兮辞》曰：“云无心以出岫”。宋人晁补之《满庭芳》词曰：“无止，流泉自急，此意本来闲。”在幽深的溪谷中，此亭一“点”，顿生人情，并将游人引向南端。同时，此亭与夏园南端的“清芬斋”形成相互的对景，在“涤虑亭”可仰望“清芬斋”，在“清芬斋”可俯临整个峪园。一俯一仰，一高一低，加之一瀑布，一流水，将两处景点连为一体，使整个峪园生机盎然。“水流云在”与“涤虑亭”两个景点，妙趣横生，意境无穷：让云气自生，让碧水自流，让心情自在。

峪园溪流荷茅亭
Ravine Garden - The thatched pavilion and the stream

Worries”. By placing this pavilion in the deep ravine, the scenery suddenly becomes vivid because it lures the visitor to the very southern end of Liu Fang Yuan. So this scenery is named “secluded ravine and clear mind”. In addition, a small building named “Studio of Pure Scents” on the southern end of the Summer Garden sits on top of the cliff and cascade. This seemingly echoes the scenery of the stream and the pavilion, one is on the top of the stream and the other at the bottom, one with a commanding view of the gully and the other with the views of the cascade and the Studio above. These two scenes of “tumbling waters merge into the upper clouds” and “secluded ravine and clear minded” make the Ravine Garden as attractive as a poem: let the cloud appear by itself; let the clear water flow by itself; let the mind be free.

秋园：万壑枫深，平湖秋月

秋园坐落于流芳园中心湖区的西面，背靠一个较陡的坡地。主要景观是由一座太湖石叠成的大假山及叠水瀑布、一座船舫、一个洲岛和两座曲桥以及秋色植物构成。计成在《园冶・相地》中云："掇石莫知山假，到桥若谓津通。"假山沿坡面叠，上有石梁，下有沟壑，高低起伏，层峦叠嶂。在假山之上设有叠水及瀑布顺山势而下，汇入湖中，形成"高山流水"之态势。园林水系易成，山势难立。清代散文家阮葵生《茶余客话》曰："以意垒石为假

Autumn Garden: myriad ravines in deep maples, autumn moon over the tranquil lake

The Autumn Garden is located on a slope on the western side of the lake in Liu Fang Yuan. It consists of a large Taihu rockery hillock or mountain with a water fall, a stone pavilion, a boat-shaped pavilion, an islet, two stone bridges as well as autumn foliage plants. The rockery hillock

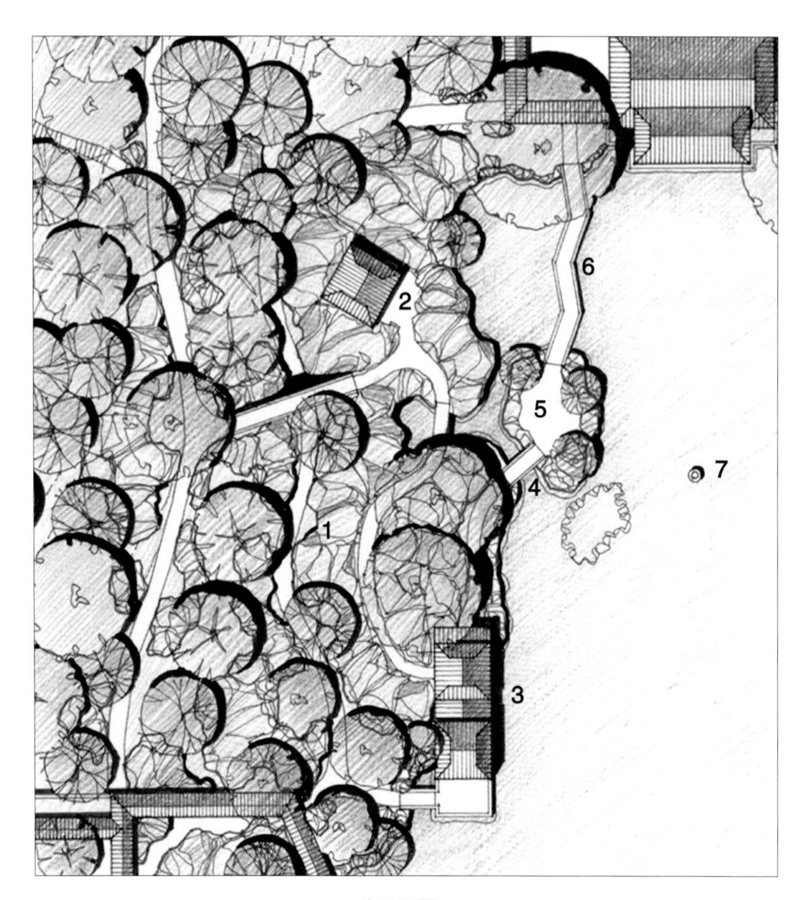

秋园平面图
Plan of Autumn Garden

1. 大假山及瀑布
   Big Rockery Mountain and Water Fall
2. 秋月亭
   Autumn Moon Pavilion
3. 画舫（与谁同坐舫）
   Boat-shaped Building
4. 步月桥
   Walking the Moon Bridge
5. 洲岛
   Island
6. 鱼乐桥
   Bridge of the Joy of Fish
7. 印月塔
   Reflecting the Moon Stone Pillar

秋园透视草图
The sketch persepective of the Autumn Garden

山。”园林假山是真山的替代，寄寓着中国人对山的情怀。假山，不是真山，胜似真山。“维假山，则又自然，真山也。”（明孙国光《游勺园记》）假山怪石，奇形百态，超俗拔尘，具有散逸淡泊，甘于寂寞，淡于功名的象征意义。假山是中国园林的一大特色。假山“假”在不是完全模仿自然之山，“真”在以自然石料而作，妙在似与不似之间。计成《园冶·掇山》云：“有真为假，做假成真。”假山需有“巧夺天工”之妙，而“假山之妙，就在虚虚实实、真真假假之间。”（朱良志）

is constructed with various forms such as stone bridges, gullies and grottoes, with ups and downs, and overlapping peaks. A waterfall descends from the top of the hillock like a mountain stream running down to the lake. In the Chinese garden, it is relatively easier to arrange water course than to pile up a hillock. Ruan Kuisheng (AD 1727—1789), a Qing dynasty essayist, said: "*Jia shan* or man-made hillock is constructed with rocks in accordance with artistic conception." The man-made hillocks in Chinese gardens are the substitute for real mountains, embodying the feelings of the Chinese about the natural mountains. The man-made rockery hillocks and wonderful rocks, which look grotesque, symbolize the human virtues of indifference to fame and embracing loneliness. The man-made rockery hillocks are one of the major features in Chinese gardens. They look like "fake" because they are not exact imitations of natural mountains, while they look like "real" because they

秋园“万壑枫涛”
Autumn Garden - myriad ravines in deep maples

根据设计，在假山之巅，设一座石制亭台，周边植几株枫香、银杏和红枫等秋叶树。每当初秋来临，气爽山寒，枫叶灿烂，色彩斑斓，而有“树叶纷纷落，乾坤报早秋”的景象。深秋降至，秋水寒山、山高水长，又有几分“荆溪白石出，天寒红叶稀”（王维《山中》）的意境。每当明月当空，登临石上，闲坐亭中，四面清风，可观赏空中明月，水中月影。故此亭取名为“秋月亭”，有元代画家曹云西自题《秋林亭子图》诗“云山淡含烟，万影弄秋色。幽人期不来，空亭倚萝薜”之意境。明代祁彪佳赏亭时说：“此亭不昵于山，故能

are constructed with natural rocks, and they are wonderful because of this combination of likeness and unlikeness . Ji Cheng said: “Some of the real look like fakes, while the man-made should be created to look real.” So it is necessary to design and construct the rockery hillock with wonderful workmanship to excel natural mountains. “The wonderfulness of man-made rockery hillock lies between likeness and unlikeness, realism and unrealism,” said Zhu Liangzhi.

In the design of the Autumn Garden, there will be a stone-made pavilion and platform sitting on top of the rockery hillock, where several foliage trees such as liquidambars, gingko trees and maples are planted around the pavilion. In a refreshing autumn weather, colorful foliages of these trees mark the arrival of the autumn. In later autumn, the weather is cool, the running water is

尽有山，几叠楼台，嵌入苍崖翠壁，时有云气，往来飘渺，掖层霄而上，仰面贪看，恍然置身无际，若并不知有亭也。倏忽回目，乃在一水中，激石穿林，泠泠传响，非但可以乐饥，且涤十年尘土肠胃。夫置屿于池，置亭于屿，如大海一沤，然而众妙都焉，安得不动高人之欣赏乎？” 正可谓“江山无限景，都聚一亭中”。秋日游园，凉风四至，月出树梢，大有“栖志岩壑，量广识明”之意“烟翠三秋色，波涛万古痕”之韵。秋园的假山、叠水、亭台尽得此境界也！假山景点取名为“万壑枫深”。

沿湖西岸设有画舫一只，取待渡起航之势，是观赏水景和游玩宴饮的佳处。船舫，亦称为画舫、旱船、船厅、藻舟等，甚至还有“不系舟”之称谓。船舫建筑是中国园林中的一种特殊建筑。它不是可行的船，却有着“船”的深层寓意。“船”或“舟”有“载人渡水”和“垂钓”的功能，但对于中国文人雅士来说，它却代表了一种精神的寄托和超越。中国山水画中，画家常常在烟雨的水面上，画扁舟待渡或渔翁垂钓，这种意境表现了野逸超凡的情怀，这种意象代表艺术家挣脱尘世，“渡向精神的彼岸”（朱良志）之意。隋唐道士成玄英在《庄子注疏》中曰：“唯圣人泛然无系，譬彼虚舟，任运逍遥。”北宋文学家欧阳修更是在《画舫斋记》中将画舫描述成“如偃休乎舟中……又似泛乎中流”。李白诗云：“人生在世不称意，明朝散发弄扁舟。”画舫还有“渔隐”之意，以求完善自己的心性修养。

cold, most leaves have fallen down, and this scenery looks very much like the artistic conception described in Tang dynasty poet Wang Wei's poem of *In the Mountains*: "The white rockery appears out of a stream and only a few scattered red leaves in the cold weather." In a clear night with the moon in the sky, climbing up to the rockery and sitting inside the pavilion, the visitor enjoys the cool breeze and the bright moon as well as the reflections in the lake. The pavilion is so named the "Autumn Moon Pavilion". Qi Biaojia (AD 1602—1645), a Ming dynasty scholar, wrote lines about a pavilion: "The pavilion gains the views of the mountain though not sitting in it; with its storied platforms, the pavilion embeds in the green cliffs; clouds and mist come and go rise and fall; one looks up and suddenly feels exposed to the bigger landscape, not knowing the existence of the pavilion. And suddenly looking back, one feels being in the middle of the water which runs through the woods and makes sounds; these make one feel no hunger and wash away the dust in the stomach. I put the islet in the day, the pavilion in the islet. It's like a bubble in the ocean. With all the miracles here, how can not the lofted gentlemen appreciate all these?" The Autumn Garden with the rockery hillock, the waterfall and the "Autumn Moon Pavilion", obtains all these exquisite scenes. The scenery of the rockery hillock is named "myriad ravines in deep maples".

There is a boat-shaped pavilion sitting on the western bank of the lake, which seems ready to set sail. It is the best place for looking at the water scenery and hosting a fete. A boat-shaped building is also called dry boat, boat-hall or algae boat. It is a unique architectural feature among

在画舫北面，一条小径蜿蜒穿过假山，另一小径沿湖北上，跨过一石拱小桥，到达一座小岛屿。临近悠悠湖上，布置一座印月石塔，点缀于湖中，游鱼环绕，波光粼粼，有邈然千里之意。古人曰："水之奇在月"。当秋夜月明时，月来云破花弄影，水上明月入园中，澄潭映空，花影弄墙，赏灯品桂，游览夜中庭院，别有意趣。陈从周先生有诗云："近水楼台先得月，临流泉石最宜秋。"

在小岛屿北面，一座五曲石桥连接到冬园。曲桥横卧波面，游鱼与波光上下，故将这座桥取名为"鱼乐桥"，这也是借战国时期道家庄子与惠子濠上鱼乐之辩的典故。《庄子·外篇·秋水》：庄子与惠子游于濠梁之上，庄子曰："鲦鱼出游从容，是鱼之乐也。"惠子曰："子非鱼，安知鱼之乐？"庄子曰："子非我，安知我不知鱼之乐？"惠子曰："我非子，固不知子矣；子固非鱼也，子之不知鱼之乐全矣！"庄子曰："请循其本。子曰'汝安知鱼乐'云者，既已知吾知之而问我。我知之濠上也。"这个景点让游人联想到这个著名的典故，诠释出庄子的"天地与我并生，万物于我为一"的"齐物"哲学思想，强调人对生命宇宙的体验和参悟，从而进入一个与天地优游的境界。但见水中鱼点点，心与之往来，享受一种游鱼之乐。此处濒临湖面，可观鱼、赏月、泛舟，主题景致取名为"平湖秋月"。

the Chinese garden buildings. It is not a real boat for traveling, but has some symbolic implications regarding the "boat" in Chinese culture. "Boat" has the functions of ferrying to the other side of the water and holding people to fish on the water. To the Chinese scholars, however, it presents the metaphors for spiritual attachment and transcendence. In most Chinese landscape paintings, the painter usually draws a small boat on the water that is carrying either a passenger or a fisherman. These scenes present the artist's longing for escaping from this mortal world to the spiritual shore or for the spiritual freedom. Cheng Xuanying (AD 608—?), a Tang dynasty Daoist said in his notes on *Zhuang Zi*: "Only the sage is free from worries, like a boat traveling as it pleases, and unfettered." Li Bai said: "If the world is in no way to satisfy me, then next day I will loosen my hair and go boating." The boat-shaped pavilion also symbolizes one's seclusion for self-cultivation.

On the northern side of the boat-shaped pavilion, a winding path goes through the rockery hillock, and another path diverts to the islet through an arched stone bridge. On the leisurely lake, a stone lantern sits in the water near the bank, where carps are swimming in the water. This assemblage creates a scene that looks mysterious and remote. The ancient Chinese said: "The wonder of water appears when the moon comes out from the sky." In a clear autumn night, the moon comes, the clouds disperse, the flowers play with their shadows shown on the white-washed walls, the sky and the moon are reflected in the waters inside the garden, , and the lantern illuminated like the moon. Meanwhile, besides appreciating the scenery, visitors can enjoy the taste the moon cakes flavored with osmanthus flowers. All these create a unique interest and charm of the Autumn Garden.

秋园“平湖秋月”

Autumn Garden - autumn moon over the tranquil lake

流芳园 － 秋色

LFY - The Autumn colors

游览流芳园之秋园，一可纵观于飞瀑之假山，二可攀援于叠嶂之洞壑，三可仰望于拂云之“秋月亭”，四可俯瞰于澄虚之“三友阁”，五可泛游于画舫之舟楫，六可鉴赏于明月之印塔。秋园的总体意境可用王维《山中》的诗意来概括：“荆溪白石出，天寒红叶稀。山路元无雨，空翠湿人衣。”

Further to the north from the islet, a zigzag stone bridge connects to the Winter Garden. This bridge reposes on the water, offering the best spot for watching these colorful carps. Its name is "Bridge of the Joy of Fish", an analogy with a famous literary quotation recorded in the debate between Zhuang Zi and Hui Zi during the Warring States period (475—221 BC) in China. In a syncopated version of the debate, Zhuang Zhi said that the fish in the water are happy, while Hui Zi said to Zguang Zi: "As you're not the fish, how do you know the fish are happy or not?" Then Zhuang Zi said to Hui Zi: "As you are not me, how do you know I don't know about the fish?". This scenery reminds the visitor that the famous legendary tail, which explains Zhuang Zi's philosophic concept of "Heaven and earth co-exist with me, and all things unify with me into a whole", with emphases on one's experience and understanding of the universe, and entering the realm of a leisurely and carefree journey in the universe. More or less, the visitor enjoys watching the colorful carps and feels as happy as the fish in the water. Around this spot by the lake, the visitor will enjoy watching fish, looking at the moon, and boating out to the water. This scenery is named "the autumn moon over the tranquil lake".

Wandering in the Autumn Garden, the visitor can enjoy many activities and varied views. The visitor may look at the waterfall from the large rockery hillock, climb and walk up and down the hillock, look up at the "Autumn Moon Pavilion", look down at the distant "Pavilion of Three Friends" from the top of the hillock, make a "boat tour" in the garden, or enjoy watching the moon reflection through the holes of the stone lantern. The overall artistic conception of the Autumn Garden is something like the one described in Tang dynasty poet Wang Wei's poem of *In the Mountains*: "The white rockery appears out of a stream and only a few scattered red leaves in the cold weather. There is no rain over the mountain path, but one gets wet while wandering within it."

流芳园 – 景墙与红叶
LFY - garden wall and red foliage

流芳园 – 叠瀑
LFY - Waterfall

冬园：文翰音清，古木风流

冬园位于流芳园的中心区，坐北朝南，临湖而建，是园中最大和最主要的庭院。这里地势平坦，数株橡树参差其间，平地北面和西面是缓坡和松林，是一块风水宝地。冬园主要有三个主题景观：一是主建筑庭院，由“翠微轩”、“同庆堂”、戏台和藏书楼围合而成的一处“文澜音清”的景区；二是由自然溪流构筑一处“曲水流觞”的景点；三是利用一株枝叶茂盛、树形独特、饶有古致的大橡树而构成的“古木风流”的景点。

Winter Garden: literary and musical realm, distinguished and admirable old tree

The Winter Garden is situated almost in the center of Liu Fang Yuan. Its main buildings and courtyards, which faces the south, are the largest in the whole garden. The site is a flat area with a few giant oak trees around and gentle slopes on its west and north. It is an auspicious spot according to *feng shui* principles. There are three major sceneries in the Winter Garden: first, the main courtyards and buildings including a waterside pavilion, an opera pavilion and a library tower

1. 翠微轩
   Cui Wei Pavilion
2. 同庆堂（清越台）
   Celebration Hall
3. 戏楼
   Opera House
4. 戏台
   Stage
5. 藏书楼（文澜楼）
   Library Hall
6. 曲水流觞
   Winding Stream
7. 兰亭
   Orchid Pavilion
8. 古木风流（大橡树）
   The Charming Old Tree

冬园平面图
Plan of Winter Garden

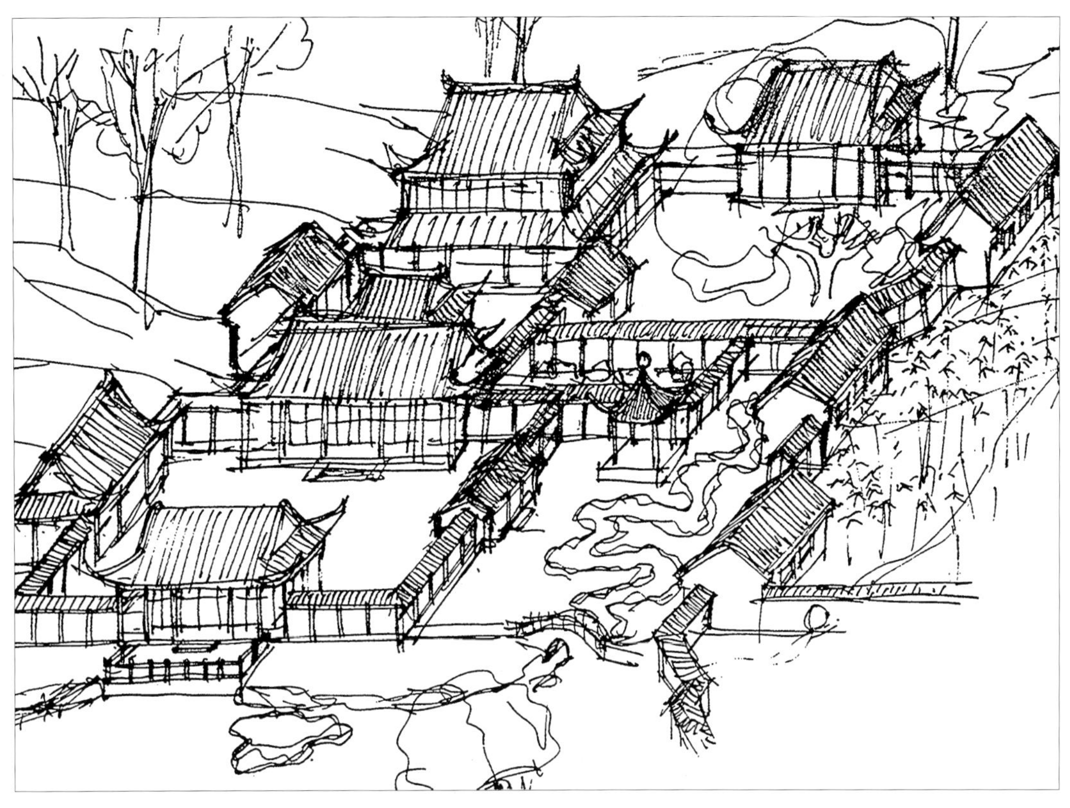

冬园透视草图
The sketch perspective of the Winter Garden

看戏，是古代文人士大夫乃至帝王生活中重要的内容，特别是昆曲和京剧，在明清时期，就融入园林当中。园林中的厅堂、水亭都可作为戏曲的表演场所，部分园林甚至在其中还特别建有戏台。戏曲融入园林，主要表现在两个方面：一是对曲的表演和观赏成为园林文化生活的一部分；二是曲境与园境彼此相通，相得益彰。陈从周先生独到地将园林美与戏曲美结合起来，提出了“曲景与园景不可分也”，“以园解曲，以曲悟园”，“曲要静听，

for storing Chinese books, which make up the scene of “literary and musical realm”; second, a winding stream representing the legendary gathering of Wang Xizhi and his scholar friends in Lan Ting in Jin dynasty, which makes up the scene of a “winding stream with floating wine cups”; and third, a giant gnarled oak tree, and the surrounding covered walkways, which make up the scene of the “distinguished and admirable old tree”.

Watching the performance of traditional Chinese opera was an important part in the life of Chinese scholars and even emperors in the past. The Kun Opera and Peking Opera were often played inside gardens in Ming and Qing dynasties. Some of the halls and waterside pavilions inside gardens held the performances of these operas, and in other gardens, there are special houses or towers for opera plays. The integration of garden with opera has two implications: one is that opera

园宜静观”等观点。明代戏曲作家汤显祖的《牡丹亭》中有曲词：“朝飞暮卷，云霞翠轩”，“雨丝风片、烟波画船”，唱出了园林的景致。明代高濂的《玉簪记》中有曲词：“粉墙花影自重重，帘卷残荷水殿风”，唱出了园林的意境。在园林之中常常设有戏台，比如，颐和园中有大戏台和小戏台两座，扬州何园中有戏台一座立于池塘之中，上海豫园也有一处戏台小院，以及苏州拙政园中的“留听阁”，网师园中的“濯缨水阁”，留园中的“明瑟楼”等，都是演戏、观戏和听戏的好地方。

藏书，也是中国文人乃至帝王的另一大爱好，是文化生活中一个不可缺少的内容。在

performance becomes part of the garden life, and the other is that the aesthetics of the opera and garden are similar to each other. For example, the setting of a garden can be seen as a stage for opera performances. Chen Congzhou had insightfully integrated the aesthetics of opera with his garden design theory and practice. He said: “The garden scenery is inseparable from the scenery of opera”, “Operas can be interpreted through gardens, and vice versa”, and “One should listen to operas quietly, the same as appreciating gardens”. In the *Peony Pavilion*, a Kun Opera written by Tang Xianzu (AD 1550—1616) of Ming dynasty, the lyrics says: “Twilights fly in the morning

冬园“文翰音清”

Winter Garden - literary and musical realm

冬园“曲水流觞”

Winter Garden - winding stream with flowing wine cups

园林中设小型的藏书楼也较为常见。汉庭顿先生本人就是一位大藏书家，汉庭顿又以其图书馆而闻名遐迩。在 2014 年，汉庭顿的学者们在馆藏的图书中又发现了中国明代的《永乐大典》的部分书卷。这些都说明“藏书”在汉庭顿具有优良的传统。所以，在流芳园的设计中，布置了一个专门收藏中国图书的二层楼阁在冬园内，这应该会对流芳园以及汉庭顿的文化内涵有所增色。此处由戏台和藏书楼组成的庭院景观区取名为“文澜音清”。

and sweep off in the evening: colorful clouds reside on top of an emerald pavilion", "rain drops and wind blows; a boat-shaped pavilion stays amidst misty waves." These words presented the garden scenery. In *A Tale of Jade Hairpin*, an opera written by Gao Lian (AD 1573—1620) of Ming dynasty, the lyrics says: "The shadows of flowers layered on the white-washed walls; rolling up the curtain, one sees the withered lotus and feels the breeze from the hall." These words presented the artistic conceptions of gardens. It is quite common to construct an opera house or pavilion inside a Chinese garden. Inside the Summer Palace, for example, there are two opera houses, one is grand and the other smaller. Inside He Garden in Yangzhou, an opera pavilion sits in the middle of a pond. Inside Yu Garden in Shanghai, there is an opera courtyard. Inside Suzhou gardens, pavilions and halls functioned as the opera stages. All these are the best places for listening and watching performances of Chinese traditional operas.

“曲水流觞”，是中国古代圣贤文人的雅聚游戏活动之一。东晋的竹林七贤，畅饮于山林水际之间，书圣王羲之会稽山兰亭的雅聚，一觞一咏，畅叙幽情。流芳园内有一条自然溪流由北向南正好穿过北部区域，即冬园的东部。设计将溪流组织成蜿蜒曲折状，模兰亭曲水流觞的形态，构成一处流动的曲水，周围点缀景石，设置亭、轩和围廊，续永和之韵，供文人雅士以及游客体验古代文人的“一觞一咏”的雅趣活动，似多吟咏幽趣，更入怀古深情。

“古木风流”的景致是由一株枝叶茂盛、树形独特、饶有古致的加州大橡树作为观赏对象，以回廊和一个主厅堂将其围合在内而成。这棵硕大的橡树，枝干发达、形状独特、虬枝拂地、幽幽蛰龙、苍古卓美、宛若画意。计成在《园冶·相地篇》中指出：“多年树木，让一步可以立根。斯谓雕栋飞楹构易，荫槐挺玉成难。”故在流芳园设计中以此树为中心构筑一个院落，供游人观赏此橡树和纳凉之用，既形成一个特色景点，又突出了当地植物的生态和观赏价值。

Book collecting is another interest for the Chinese scholars and emperors. It is a part of the cultural life in Chinese gardens. It is not rare to find small libraries inside Chinese gardens. Mr. Huntington was a preeminent book collector. The Huntington Library is an internationally recognized library. In 2014, the Huntington scholars found original copies of a section of the Ming dynasty's *Yong Le Da Dian* inside the Huntington Library. All these mean that book collecting is a tradition at the Huntington. Therefore, in Liu Fang Yuan, the design has proposed a small two-story library building inside the Winter Garden for collecting related Chinese books. This feature will enrich the cultural activities at Liu Fang Yuan as well as the Huntington. The combination of the opera pavilion and the library building will make a scene named “literary and musical realm”.

In ancient China, it was a custom that scholars seasonally got together at an exquisite landscape area to play and write poems. One of the gathering places had a stream running through, and the people sat along the stream and put wine cups in the stream to let them flow down. When a cup flows close to or stops at front of a scholar, he should pick it up, drink the wine and write a line or a poem. Whoever could not write properly, he would be fined to drink more wine. This game became a legendary activity among scholars since the famous gathering of the calligraphy master Wang Xizhi and his fellows in Lan Ting of Shaoxing in Eastern Jin dynasty. The setting then becomes a distinct garden or landscape feature, which is named the *qu shui liu shang* or “winding stream with flowing wine cups”. At the Huntington ground, there is a stream running through the area of the Winter Garden in Liu Fang Yuan from north to south. The design made a feature of “winding stream with flowing wine cups” which is surrounded with pavilions and covered walkways, in order to represent this traditional scholar activity and allow visitors to experience its charm and fun.

冬园“风流古木”

Winter Garden - distinguished and admirable old tree

The scene of the "distinguished and admirable old tree" consists of an ancient-looking, vivid and gnarled California oak tree, a main hall and an enclosed verandah. Ji Cheng said: "An old tree should be preserved because it will become an anchor or root of the garden. As it is said, it is easier to construct buildings with carved beams and tablets than to have the shade of a Chinese scholar tree and a good stance of magnolia." This California oak tree is still vigorous and beautiful with its convoluted trunks and branches. The design aims to preserve the big oak tree and the others to make the garden sustainable with the local climate and create distinguished scenery so that the visitor will have an opportunity to appreciate the ecological and ornamental values of the local trees and to learn about local plants.

宝塔园：湖山真意，沧浪环碧

在中国上古时期，有建高台以通神明的观念，楼阁就是由此而发展出来的一种园林建筑。“宝塔”，自汉代随佛教从印度传入中国后逐渐世俗化，与中国传统建筑相结合而形成了中国式的塔式建筑，并且衍生出多种形式，如楼阁式塔、密檐式塔、亭阁式塔等，建筑层数多为五、七或九层，平面形式以正方、六边和八边为多，材质有夯土、砖石、木等。塔的功能除了设在寺庙陵园内的宗教纪念性作用之外，多数是作为点景和登高望远之用，成为一个区域或城镇的标志性建筑。塔式建筑成为独具中国特色的一种传统建筑，中国的大江南北都有许多宝塔建筑，最为人所熟知的有杭州的六和塔，苏州的虎丘塔和北寺塔，北京北海的白塔，镇江的金山塔等。

在流芳园内设一座七层楼阁式宝塔，矗立于半山半水之间，可引景引人，点活整个园林。宝塔秀出于云表，倒映于水中，构成了仰视、俯视两个借景。秀挺的宝塔从整个园中不同的角度观赏都有“步移景异”的变化和风姿，处处随人，引人流连忘返。布置宝塔式建筑主要有三方面的立意：一是整个园林面积较大，建筑和庭院丰富多样，

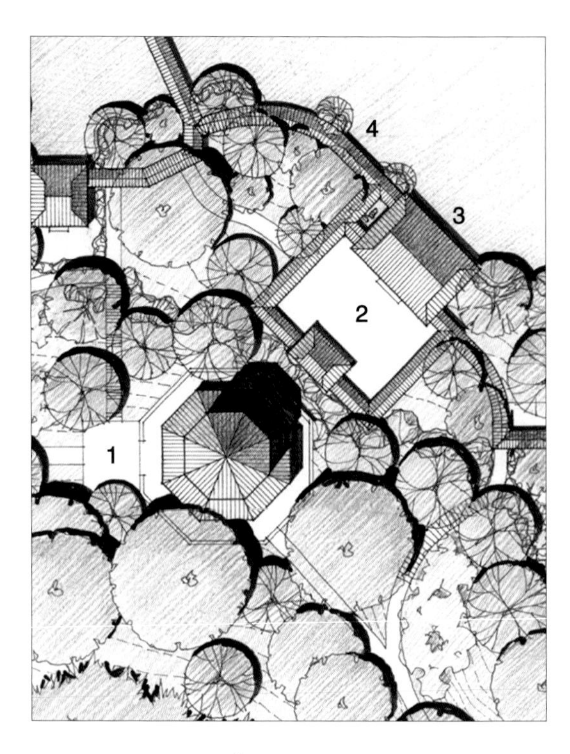

宝塔园平面图
Plan of Pagoda Garden

1. 宝塔（湖山真意）
   Pagoda
2. 四合院
   Courtyard
3. 水榭
   Waterside Pavilion
4. 曲廊（沧浪环碧）
   Zigzag Corridor

Pagoda Garden: Seeking true meaning amidst lake and mountains, azure water surrounded with bluish-green

In ancient China, tall platforms and pavilions were constructed for the purpose of making connections to the Heaven and worshiping gods. The tall pavilion or *lou ge* became a garden feature for climbing up to view the distant scenes. Pagoda or ta originated in India as a Buddhism feature. It has been secularized and transformed with Chinese culture and architecture since it was adapted to China during Han dynasty. The Chinese pagodas have various configurations such as storied pagoda, multi-eaves pagoda and pavilion pagoda, etc. They are usually five or seven or nine stories tall in square, hexagon or octagon shape. They are often made of earth, bricks, wood or mixed materials. Pagoda is a unique Chinese architectural feature not only in temples for religion purpose,

但高度均不超过二层。若有一座主体建筑统领各个建筑群落，势必应在高度上较为突出。“楼阁碍云霞而出没”，而中国的塔式建筑当然是理想的选择。二是整个园林及周边地形高低起伏，北面远处有圣盖博群山，若创造一处观赏全园和眺望远景的至高点，可使游人在游览流芳园时产生一个高潮点。山水之际，高塔斯起；野翠天碧，玲珑错落。中国文化强调“登高望远”和“更上一层楼”的境界。明人孙国光有曰：“意所畅，穷目；目所畅，穷趾。”中国园林强调“登临之乐，览者各自得焉”，追求“山楼凭远，纵目皆然”的意境。三是为流芳园创造一个突出的“点景”和“借景”的功能。宝塔秀出于山岗，塔影倒插于碧波之中，无论园内园外，皆可引人引景。结合以上三点考虑，设计将这座七层楼阁式宝塔布置在园内的西南边的坡地上，塔的内部设有楼梯和电梯，供游人登临塔顶，俯瞰全园亭台池沼，眺望远山峰峦，体会“湖山真意”，定会令游人惊叹！

在宝塔所处的山坡与中心湖畔之间的区域内，沿湖布置了一个四合院和曲廊，以连接夏园和秋园。四合院主要作为进出宝塔景点的人流集散和休息空间，在临湖一端布置一座水榭悬挑于湖上，两端以曲廊分别与夏园和秋园相连。曲栏俯清流，晴空摇翠浪。游人或品茗于水榭之中，或信步于环廊之内，环顾湖上景色，碧水蓝天，花木苍翠，亭榭曲桥，尽收眼底。此处景点取名为“沧浪环碧”。

but also in towns, landscapes and gardens for decorative purpose. The well known examples are the Pagoda of Six Harmonies in Hangzhou, the Tiger Hill Pagoda in Suzhou, the White Pagoda in Beihai Park in Beijing, and the Golden Mountain Pagoda in Zhenjiang.

Inside Liu Fang Yuan, a seven-story pagoda was designed and situated in the middle of a hillock near the lake. It will make an attractive scene and lure the visitor to the garden. It will make Liu Fang Yuan much more vivid and exquisite. The pagoda protrudes into the clouds and reflects in the waters, which make up upward viewing and a downward viewing of the scenery. Looking at the pagoda from different areas and angles within the whole garden will create an effect of so called "scenes changing in every step", scenes follow the visitor and make one linger in the garden. There are three purposes to design a pagoda inside Liu Fang Yuan: first, the total area of the whole garden is relatively large made up of many courtyards and buildings as well as landscape features. But the buildings and pavilions are all one or two stories tall. Hence, if there is a dominant structure in the whole garden, it should be taller than all other ones. The ideal feature for this purpose is a tall pagoda. Secondly, the topography inside the garden as well as the surrounding areas was ups and downs. In addition, the San Gabriel Mountains on the far north of the Huntington are magnificent. If a higher viewing point is created, it will allow visitors to enjoy a panoramic view of the whole Liu Fang Yuan and vistas of the San Gabriel Mountains and more. This spot will become the apex of the tour in the whole garden. Climbing to the top and seeing distant landscape as well as going up to a higher level is a Chinese cultural phenomenon. Sun Guoguang, a Ming dynasty scholar, said: "The mind is freed to the limit of how far the eye can see; the eye is freed to the limit of how far the feet can walk." The pleasure of climbing up higher and viewing whatever one can see is desirable in Chinese gardens. Thirdly, the design of the pagoda creates the prominent scenery which can be borrowed by the whole garden, either from different courtyards, the hillocks and the waters. Inside the seven-story pagoda, the visitor can access to each floor and the top through elevators or stairs. Once arriving at the top floor, the visitor will see the whole garden and distant views of the Huntington as well as the San Gabriel Mountains. This scenery is so named the "seeking true meaning amidst lake and mountains". What a striking and spectacular panoramic view of the garden and beyond!

Between the Pagoda and the lake shore, there is a courtyard designed for visitors to get in and out of the Pagoda and rest in this area. A waterside pavilion and covered walkways make up this courtyard, which connects to the Summer Garden and the Autumn Garden. The visitor can either drink teas inside the pavilion or stroll in the corridors to see the garden landscape: the lake water, bridges, rockery, trees and flowers, etc. This scenery is named "azure water surrounded with bluish-green".

宝塔园“沧浪环碧”

Pagoda Garden - azure water surrounded with bluish-green

盆景园：壶天揽胜，飞虹流霞

位于中心湖西边的山坡地上，设计了一座中国盆景园。在原有地形上稍加改造便可布置回廊、展室、景墙和观景亭，供摆设盆景，让游人观赏。盆景，源于中国，产生于唐代，成熟于宋代，兴盛于明清。唐朝时期，盆景由简单的植物"盆栽"升华为具有意境的"盆景"，一种以简约方式浓缩山水风光的艺术。白居易《莲石》诗句："青石一两片，白莲三四枝。寄将东洛去，必与物相随。"杜甫《天宝初》诗句："一匮功盈尺，山峰竞出群。望中凝在野，幽处欲生云。"自宋代开始，中国哲人便将盆景当作"观天地生物气象"的手段之

Pengjing Garden: seeing all the splendors in a gourd-like world, flying rainbow and flowing red clouds

On the western side of the central lake, a Penjing Garden was planned for collecting penjings or potted landscape and displaying them. The area is on a slope, which makes the space more interesting with the arrangement of garden walls, covered walkways, pavilions as well as landscape features such as spring water cascade, Taihu rocks, trees and shrubs. The art of penjing originated in Tang dynasty of China, developed in Song dynasty and thrived in Ming and Qing dynasties.

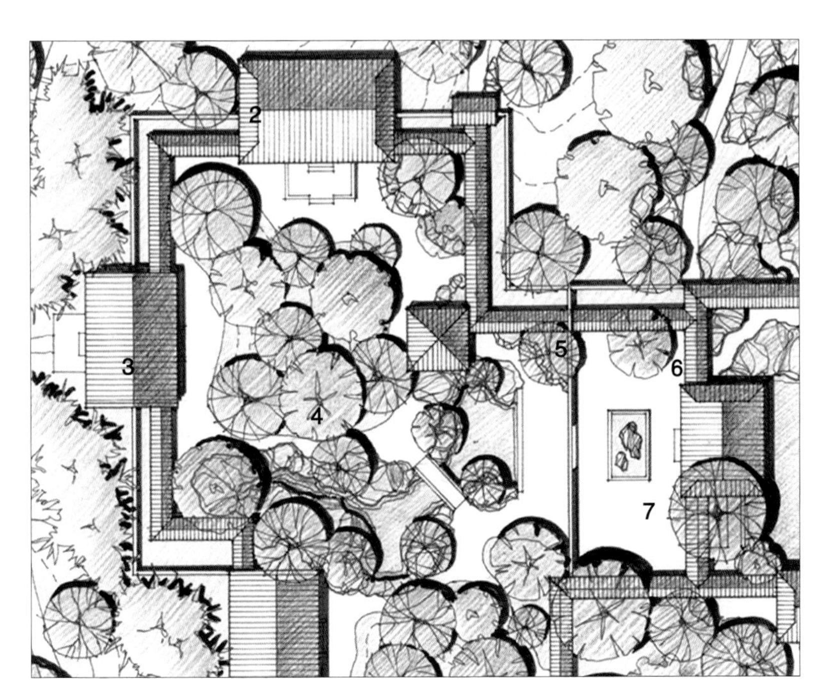

1. "壶天揽胜"庭
   Courtyard of Hu Tian Lan Sheng
2. 盆景展厅
   Penjing Exhibition Hall
3. 盆景展廊
   Penjing Gallery
4. 涌泉叠水
   Spring and Cascade
5. 天趣亭
   Pavilion of the Heavenly Interests
6. 水庭
   Water Courtyard
7. 牡丹亭
   Peony Pavilion

盆景园平面图
Plan of Penjing Garden

宝塔园“湖山真意”

Pagoda Garden - true significance of lake and mountain

一。南宋哲学家朱熹《汲清泉奇石小诗》曰："清窗出寸碧，倒影媚中川。云气一吞吐，湖江心渺然。"明代盆景理论家吕初泰在《盆景》一文中曰："盆景清芬，庭中雅趣"，"生韵生情，襟怀不恶"。

中国盆景艺术，运用"缩龙成寸"、"以小见大"的艺术手法，既顺乎自然之理，又巧夺自然之神工；既有古朴嶙峋、葱翠劲健之气势，又有潇洒清秀、艳姿丰实之仪态。中国盆景四时可赏，各具风韵，尽显天趣。朱良志先生在《真水无香・盆景的生意》一文中说："盆景是中国人艺术人生态度的活的形式"，又说："盆景就是人'生命的雕刻'，艺术家创造一片活的宇宙，是为了展现玲珑活络的心灵。"中国盆景被誉为"无声的诗，立体的画"。宋代赵希鹄《洞天清路・怪石辨》曰："石小而起峰，岩岫耸秀，嵚崟之状，可登几案观玩，

流芳园 – 彩霞映碧湖
LFY - Pink clouds reflected in the lake water

During the Tang period, penjing was transformed from potted planting or *pen zai* to artistic potted scenery or *pen jing*, an art of miniaturizing and abstracting the landscape into a pot-sized space. Bai Juyi wrote lines: "One or two pieces of rock, three or four white lotus, mailed to my friends in eastern Luoyang along with my missing of them." Du Fu wrote lines: "In a basket of several feet size, mountain peaks are competing with one another. They look wild and natural; their deep gullies seem to generate clouds." In Song period, scholars and thinkers regarded penjing as a medium to observe the cosmos and the ecological world. Zhu Xi, a Song dynasty philosopher, said in a poem: "A clear window frames the bluish-greens just inches wide and contains the charming mountains; while the air and cloud is approaching, the feeling of lake and river becomes vague." Lu Chutai, a penjing theorist of Ming dynasty, said in *Penjings*: "Refined and artistic penjings make the courtyard elegant" and "they create charms and sentiments, but no evil mind".

The Chinese penjing is created in accordance with the principles and techniques of miniaturizing and "seeing the large from the small". They are not merely imitations of the natural landscape, but also something excelling the natural appearance through the wonderful workmanship. They can be appreciated in all four seasons with different appearances in each season that present their attractions. Zhu Liangzhi said in his book of the *Real Water Without Fragrance*: "Penjing is a lively and vivid form which represents the Chinese attitude to life. Penjing is exactly

盆景园透视草图

The sketch perspective of Penjing Garden

盆景园“壶天揽胜”

Penjing Garden - seeing splendid in a gourd-like world

亦奇物也。” 苏东坡在《取弹子石养石》中吟曰：“我持此石归，袖中有东海。置之盆盎中，日与山海对。”

盆景，也是中国园林中的一种清供，布置在庭院中，摆放在厅堂内，使园林充满生机。在一些园林内还特别开辟有盆景园。流芳园内设计了一个盆景园，庭院内利用地形高差，布置以曲廊和爬山廊环绕，设几片景墙，一处涌泉，一流叠水，一勺池塘，山石花木，衬托盆景展示。空间小中见大，曲折有致，趣味横生，形成一处“壶天揽胜”的景点。游览盆景园，使人尽得风月情怀，山水情致；不出户庭，直际天地。盆景园将使流芳园更增添几分画意，几分诗情，几分天趣。

盆景园东面与中心湖区相连，临水布置一高台，其上有一亭，取名为“天趣亭”。此亭与盆景园入口景墙遥相呼应。高台之东凿一水湾，与湖相连，形成一个“水庭”，环以亭榭和廊桥。廊桥似一道飞虹跨水而过，在夕阳映照之下，如一道彩虹与晚霞浑然一体。此处景点取名为“飞虹流霞”。长虹卧碧，斜阳低尽柳如烟，好一派晚霞奇景。

like a ‘sculpture of human life’. Penjing artist creates a vivid cosmos to represent one’s spiritual world.” To the Chinese, penjing is called a silent poem, a three-dimensional painting. Zhao Xihu (AD 1231—?), a Song dynasty scholar said: “A small rock with peaks and caves looks attractive. It can be placed on a table for appreciation. What a wonderful thing!” Su Shi said: “I came back home with a gained rock inside my sleeve as if I was carrying the eastern ocean with me. Then I placed the rock inside a pot so I could look at it as if I was facing the ocean.”

Penjing is an elegant ornament in Chinese gardens, usually placed inside a courtyard or a hall. It makes the garden full of life. In some gardens, a special area is dedicated to pengjin displays or serves as a penjing garden. In the master planning of Liu Fang Yuan, a Penjing Garden is designed to take advantage of the hilly topography on the western side of the garden. It consists of zigzag verandahs and climbing corridors, a few display walls, a spring and cascade, a pond and rockery as well as trees and shrubs. All these will make the display of penjings more interesting and enchanting. The garden is just like a large penjing itself. So this scenery is named “seeing all the splendors in a gourd-like world”. Ambling inside of Penjing Garden, the visitor will have all the sentiments about the moon, the wind, the mountains and waters. Without stepping to out of the courtyard, one will experience a fantastic world. The Penjing Garden will complement Liu Fang Yuan with a bit more picturesque, poetic and naturalistic attractions.

On the eastern side of Penjing Garden, where the garden meets the central lake, there is a tall platform and a pavilion on top of it. This pavilion is named "Pavilion of the Heavenly Interests", which is complementary with the entry gate to the Penjing Garden. On the east of the platform, an inlet of water connects to the lake, surrounded by pavilions and a covered bridge, which becomes the "Water Court". The covered bridge across the water looks like a rainbow in a nice sunset glow. This scenery is named the "flying rainbow and flowing red clouds". In a clear evening, the colorful clouds of sunset reflects in the waters, the covered bridge is illuminated like a rainbow, and the twilight bathes the misty willow trees, all of which make a wonderful and unforgettable scene in the garden.

盆景园“飞虹流霞”

Penjing Garden - flying rainbow and flowing red clouds

幽竹园：竹径通幽，曲溪探梅

在流芳园的北部靠东的区域，原来有一片小竹林，设计将这片竹林保留并扩大，构建一座幽竹园，种植较为高大的毛竹，再布置一条曲径穿过其中，沿途设一竹亭，寓有王维《竹里馆》中“独坐幽篁里，弹琴复长啸”的诗情和禅意。“竹”的比德寓意也是中国历代文人墨客所喜爱颂扬的。比如，“竹节”，比喻高风亮节；“竹青”，代表四季常青；“竹空”，象征谦虚，亦佛教的“空”与“无”的含义，等等。宋代苏轼酷爱竹，他有著名的“宁使食无肉，不可居无竹”之说。竹还有四美：色泽美、姿态美、音韵美、意境美。人处其中，

Bamboo Garden: path in bamboo grove leading to the hidden; searching out plum along a crooked stream

There is an existing bamboo grove located on the northeastern part of Liu Fang Yuan. In the design, these bamboos are preserved for a future Bamboo Garden. In addition, more large bamboos such as *mao zhu* or moso bamboo will be planted. A bowered winding path and a bamboo pavilion are placed inside the bamboo grove. This garden meant to be a fresh and quiet place for the visitor to refresh and rest. The artistic conception for this scenery comes from the poetic and Zen meaning described in Wang Wei's poem: "Sitting alone in the bamboo grove, I am playing my lute and humming songs." The symbolic implications of bamboo are extolled by the Chinese scholars throughout the history. Bamboo nodes, for instance, stand for human integrity, its green for vitality, its hollow stem for rectitude and humility, etc. Su Shi of Song dynasty made a famous saying of bamboo: "I rather eat without meat than live without bamboo." There are also four different aspects of beauty associated with bamboo: the beauties of its color, stance, sound and poetic charm. The scenery in the Bamboo Garden is named "path in bamboo grove leading to the hidden".

Deep inside the bamboo grove, there is a small pavilion made of bamboo stems. This feature is distinct and attractive. One of Bai Juyi's poems said: "Relaxation doesn't need to go out far away; only a small pavilion with a few yards space will

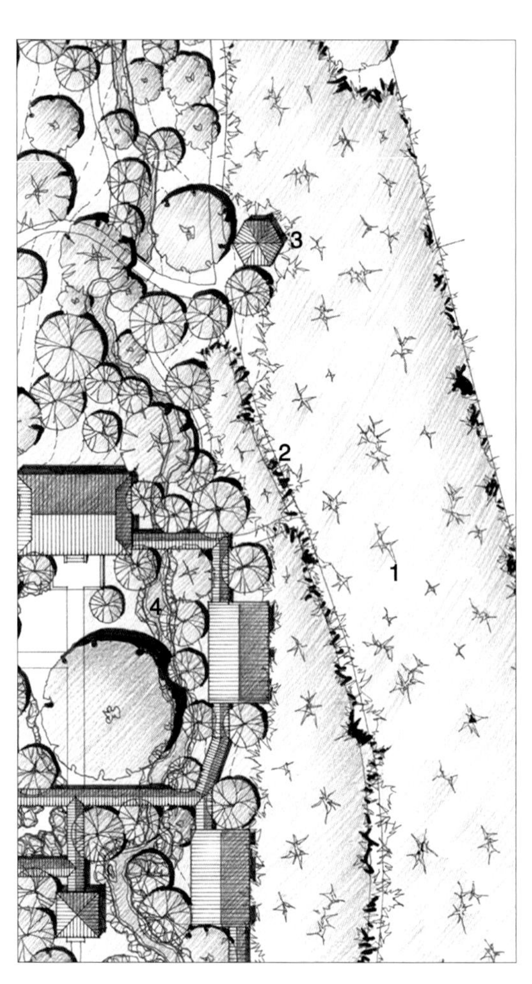

幽竹园平面图
Plan of Bamboo Garden

1. 竹林
   Bamboo Grove
2. 曲径
   Windering Path
3. 竹亭
   Bamboo Pavilion
4. 曲溪
   Windering Stream

幽竹园“竹径通幽”
Bamboo Garden - path in bamboo grove leading to the hidden

清香扑鼻，青色养眼，沁人心脾。此处景点取名为“竹径通幽”。游人沿小径北上，可探寻梅花的暗香和疏影。此处构成了松、竹、梅岁寒三友相聚成趣的景致，在当地的冬季，可体会三友的傲霜之风骨，耐寒之气节，尽显“真岁寒之丽姿”（恽南田）。此处景点取名为“曲溪探梅”。

在幽深的竹林当中设置一座小亭，自有别致的幽趣。白居易《病假中南亭闲望》曰：“闲意不在远，小亭方丈间。西檐竹梢上，坐见太白山。”元代画家吴镇擅长画竹，他有诗曰：

make it. Above the western eave sits on top of the bamboo, I can see Taibai Mountains while sitting inside the pavilion.” Wu Zhen (AD 1280—1354), a well-known bamboo painter of Yuan dynasty, said: “I have a pavilion deep inside the bamboo grove. I listen to the autumn sound to know when shall I leave or come back.” A pavilion in Chinese garden functions as a focus of view, a view finder and a vista. It is like the key lines in poetry or the finishing touch in painting. Zhu Liangzhi said in the book of the *Spirit of Life in Chinese Arts*: “Placing a pavilion is to awake and please one’s heart, enhance one’s spirituality. Clouds roll back and forth; mist permeates the vast and mighty space; all these are attributed to the pavilion.” Ji Cheng was also fond of bamboo. He said in *Yuan Ye*: “Searching for tranquility and peace in a bamboo grove, I am obsessed with it.” In the sequestered Bamboo Garden, the endless bamboo green, the tall and thin bamboos, the scenery are all beautiful;

"我亦有亭深竹里，也思归去听秋声。" "亭" 在园林中往往起到点景、引景和对景的作用。好似诗中的"诗眼"，画中的"点睛"之处。朱良志先生在《中国艺术的生命精神》中说："一亭之设，可以醒心惬意，提升性灵。云气舒卷，烽烟浩荡，皆归于一亭。" 计成也视翠竹为园林里不可或缺的景观，他说："竹坞寻幽，醉心即是。"在竹园里，翠竹幽深，修竹承露，景色宵妍，怎么不会令游客心醉于这幽竹园呢?

竹林北面，一条小溪流过，两旁松树林立，在林下种植腊梅，在溪旁布置小径蜿蜒曲折，跨溪而过。游人沿小径北上，可探寻梅花的暗香和疏影。此处构成了松、竹、梅岁寒三友相聚成趣的景致，在当地的冬季，可体会三友的傲霜之风骨，耐寒之气节，尽显"真岁寒之丽姿"。此处景点取名为"曲溪探梅"。

how can not the visitor be enchanted with the Bamboo Garden?

On the north of the bamboo grove, a crooked stream runs through an area of pine trees and down to the Bamboo Garden. In the design, wintersweet trees will be planted along the stream in this area, and winding paths are arranged to cross the stream several times and connect with the rest of the garden. Strolling on these paths, the visitor will smell the subtle fragrance of the wintersweet flowers and see the scattered shadows of the trees. The prominent plants in this area are pine trees, bamboos and plum trees, which make up the "three friends of winter" scenery. During the winter time in this region, though not too cold and possibly no snow, the visitor will still experience the charm of these "three friends of winter". So, this scenery is named "searching out plum along a crooked stream".

幽竹园"曲溪探梅"
Bamboo Garden - searching out plum along a crooked stream

松涛园：松风涧水、积翠听涛

最北面这里是全园内最疏朗的一片林地，以松树为主，园之东面有一溪涧自北向南流过。设计中将此处构建成松涛园，布置一亭、一轩于其中，二三条小径穿越其间。这里的景观无需大动干戈，正如计成所谓“略成小筑，足征大观也”。这里是“畅情林木”的好去处，夏季的清风吹拂着松枝，发出的声音仿佛浪涛一样，与溪涧的潺潺流水声交织成趣，真有王维的“明月松间照，清泉石上流”的意境。

Pines Garden: pine breeze and mountain stream; listen to the waves of pines in the emerald forest

The far northern part of Liu Fang Yuan is the most spacious area with scattered large pine trees inside the whole garden. The stream courses through the landscape on the eastern side. In the master plan, this area serves as the Pines Garden, consisting of a pavilion, a gazebo, and a couple of paths as well as the stream. The Pines Garden does not need much intervention by human hand.

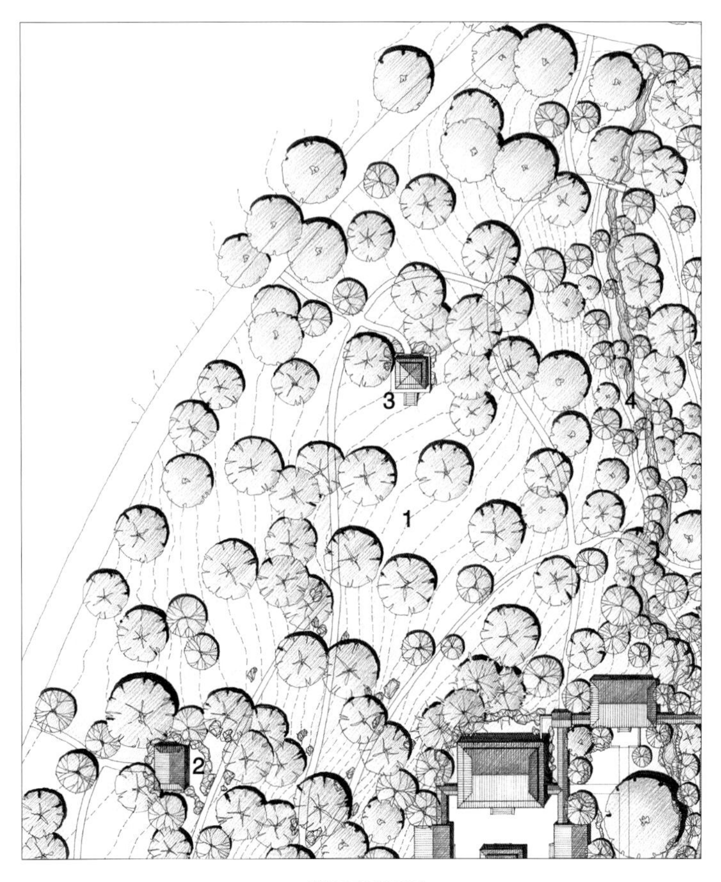

1. 松林
   Pine Forest
2. 听涛轩
   Ting Tao Pavilion
3. 松风积翠亭
   Pine Breeze Pavilion
4. 曲溪
   Windering Stream

松涛园平面图
Plan of Pine Garden

这片疏林地与流芳园其他建筑庭院形成鲜明的空间对比，一疏一密、一旷一奥、一大一小。如果将流芳园看作是一幅巨型的山水画，其空间格局好似中国画中的“疏可走马，密不容针”的布局特点。松涛园内，疏旷野逸，苍岩危立，松林幽涧，松阴晨润，山气夕爽，风起涛鸣，浩然天成。明代画家沈周自题《山水四幅》云：“行尽崎岖路万盘，满山空翠湿衣寒；松风涧水天然调，抱得琴来不用弹。”道出了“松风涧水、积翠听涛”的意境。

综上所述，流芳园的“九园十八景”，是整个园林分为九个主要景观区和十八个主题景点的设计构思，它们形成了流芳园的独特景观，整个设计是密切结合场地的地形地貌特征而产生出来的，并且突出了“园以景胜，景以园异”的中国造园思想以及传统园林的造园法则。

整个流芳园布置幽雅多致，内容丰富，并且构成一个完美的整体。中列山水，四周环宇，楼阁廊屋，高低错落，迤逦相续，深溪幽壑，势若天成。厅堂依山架构，亭榭临湖高下，廊桥随势曲折；山林起伏，以相映望；湖光泉流，相互穿透；奇石丘壑，古朴苍雄；佳木秀草，翠荫芳香；游鳞翔羽，自相映带；烟云蓊郁，晨夕万状；衣香人影，山水幽情；极天然之趣，自成清华。整个流芳园可谓“处处有情，面面生意”，能让观者“顿开尘外想，拟入画中行”。

It will make up itself with the pine trees and the terrain, as Ji Cheng said in *Yuan Ye*: “Strategically place a couple of small structures; it will be sufficient to make significant views.” Here is the place for enjoying the woods, where the pines whisper in breezes like the sound of ocean waves, and the stream runs like babbling brooks. These will make poetic scenery similar to the artistic conception described in Wang Wei’s poem: “The bright moon shines through the pine trees; the clear spring flows over the rocks.”

The spacious Pines Garden is in contrast with the ones that consist of more buildings and courtyards: one is sparse, the other dense; one is open, the other obscure; and one is large, the other small. If Liu Fang Yuan is compared to a huge landscape painting, its spatial organization follows the concept in Chinese landscape painting that “the wide open space is big enough for horse running, while the small space is dense enough for no needles to put in”. Inside the Pine Garden, the sparse and open spaces look wild, aged tall trees stand upward to the sky, stream runs between pine trees, the shade of pines turns misty in the morning, the air in the hillock is clear in the evening, and the breeze passing the pine trees makes sound like waves; all these look natural. Shen Zhou (AD 1427—1509) a Ming dynasty painter, wrote inscriptions which said: “I travel to the end on the winding paths; the mountain greens make clothes wet and cold; pine breeze and stream make sounds like natural music, so I do not need to play the zither.” These are exact artistic conceptions of the scenery in the Pine Garden: “pine breeze and mountain stream” and “listening to the waves of emerald pines”.

In summary, the "Nine Gardens and Eighteen Views" in Liu Fang Yuan are conceptualized and designed specifically, responding to the unique characters of the site and the Huntington environment. They become the distinct and attractive assemblage of Liu Fang Yuan. The overall design is a result of the applications of the concepts and principles of Chinese garden to the particular site. The design has complied with the concept that a garden will be superior with its sceneries and views; the sceneries and views should be distinct in different gardens.

The overall layout of Liu Fang Yuan is sophisticated and compelling. All the garden elements are integrated into a dynamically balanced whole. The garden is centered with a lake surrounded with buildings and courtyards, hillocks, pavilions and walkways with ups and downs, deep streams and secluded gullies; all these elements become something naturally created and "grown" out of the site. The halls are constructed next to the hillocks; pavilions and gazebos sit by the waters; walkways and bridges made turns in accordance with the site and space; hillocks and woods are corresponding to each other; lake water and stream waters are intertwined; mists and sunsets come and go; fragrances perfume the air and the clothes of visitors; mountains and waters make exquisite feelings; all these are enchanting, elegant and transcendently beautiful. Liu Fang Yuan is a garden that generates emotions, inspirations and imaginations. In the garden, the assemblage seems to make visitors imagine beyond the confines of this day-to-day world and feel as though they are wandering within a landscape painting.

松涛园“积翠听涛”

Pine Garden - listen to the waves of pines in the emerald forest

一轮明月挂树梢，三孔石桥横镜湖 The silvery moon hangs on top of the trees, the three-arch bridge lies in the smooth reflecting lake.

流芳园 - 玉带桥

LFY - the Jade-ribbonBridge

第六篇

诗情画意

Chapter Six

The Poetic and Picturesque Charm of the Garden

诗情画意

The Poetic and Picturesque Charm of the Garden

中国古代是一个诗的国度！中国古代的圣贤哲人、文人雅士曾以诗意的态度来看待人生、社会和自然。中国文化和艺术离不开诗歌，中国文人、书画家、造园家、戏曲家也离不开诗歌！诗言志，词抒情，诗词写景是为传情达意。宋代大文学家苏轼在其《东坡题跋·书摩诘〈蓝关烟雨图〉》中评论唐大诗人和画家王维的作品时说："味摩诘之诗，诗中有画；观摩诘之画，画中有诗。"这是对中国诗画艺术最精辟的诠释之一，也道出了"诗情画意"的真意。王维《山居秋暝》诗："空山新雨后，天气晚来秋。明月松间照，清泉石上流。"苏轼《涵虚亭》诗："水轩花榭两争妍，秋月春风各自偏。惟有山亭无一物，坐观万景得天全"。这些诗词描绘了"景"，表达了"意"，抒发了"情"。

The ancient China was a kingdom of poetry! The Chinese sages and philosophers, scholars and literate people, all adopted the poetic manner to see the life, society and nature. The Chinese culture and arts are inseparable from poetry, so as the scholars, calligraphers and painters, garden designers and traditional opera artists. Poetry expresses one's ideals and emotions. The scenes described in poetry convey the poets' sentiments and conceptions. Su Shi extolled Wang Wei (Wang Mojie) writing: "When one savors Mojie's poems, there are paintings in them; while one looks at Mojie's paintings, there are poems." This is the most incisive annotation on the art of Chinese poetry and painting. It also conveys the true essence of poetic and picturesque charm. Wang Wei wrote verse in the *Mountain Dwelling in Autumn Evening*: "Right after a rain in the empty mountain, the late autumn weather in the evening, the bright moon shines through the pine trees, the clear spring runs over the rocks." Su Shi wrote verse in *Waterside Pavilion*: "A waterside gazebo competes its beauty with a pavilion surrounded by flowers; the autumn moon and spring wind has their own favors. Only the pavilion does not have its own things, but sitting inside the pavilion; one looks around and then sees countless scenes of the world." These poems described the scenes, conveyed the meanings, and expressed their emotions.

One of the prominent characteristics of Chinese arts is that they are interconnected and referenced each other. It is the case in poetry and painting as well as in garden design and construction. To talk about Chinese gardens is inevitable in talking about the Chinese poetry and

静听松风图轴　宋　马麟　台北故宫博物院藏
Ma Lin (1180-1256), Listening to the Wind in Pines, National Palace Museum

中国艺术的一大特征就是可以相互贯通和相互借鉴。诗画如此，造园亦如此。谈论中国园林总离不开中国诗文和绘画。中国园林艺术非常讲求情景交融，有人、有情、有景，园林才会有生命，才会美，才会令人陶醉。诗词是以“言”的方式将“景”与“情”融为一体，表达出思想感情的“境界”。园林以诗文意境促进了“人”与“园”的联系，加强了“情”与“景”的融合，表现了造园主人的思想和情趣，提升了观赏者的艺术和精神境界。明代书画家陈继儒《青莲山房》诗曰："主人无俗态，筑圃见文心。”陈从周先生认为，“中国园林，名之为‘文人园’，它是饶有书卷气的园林艺术”。所以，中国园林艺术必然讲求“诗情”与“画意”。

中国园林中的诗情画意不仅以景致来表达，而且以题咏楹联匾额的方式来“画龙点睛”和提示“象外”之意境。中国古人将诗、书、画称之为文人“三绝”，而在绘画上题诗作文则始于明代。这种艺术形式将诗、书、画融为一体，足不出户，情怀尽抒于纸上，山水尽收于眼底，意境尽得于心中。

painting. The art of Chinese garden pays particular attention to the integration of sentiment with scenery. Inside a garden, people, emotions and scenes make the garden vivid, beautiful and enchanting. The poetry makes the integration of sentiment with scenery by words in order to express the higher state of thoughts and ideals. The relationship between human being and garden is enhanced by poetic expressions of the garden scenery. In addition, poetic expressions present the ideology and sentiment of the owner and designer of the garden, and at the same time, elevate the viewers artistic and spiritual realms. Chen Jiru (AD 1558—1639), a Ming dynasty painter, wrote verse in the *Hillock House of Green Lotus*: “If the owner and designer have no vulgar tastes, their literary and artistic thoughts can be seen from the garden construction.” In Chen Congzhou's point of view: “The Chinese garden can be called the scholar garden because of its rich literary connotations.” The art of Chinese garden, therefore, must stress the poetic as well as picturesque charms in the garden design and construction.

明清时期，园林题咏成为造园不可或缺的一部分，是一种提升园林景观意境的艺术手段，也是一项增加造园趣味的艺术活动。其实，园林中的题咏，一是点出意境，二是装饰园林，三是启发观者再创其心中的新意。以诗意造园，以园境写诗，诗即是园，园即是诗。中国园林不仅是一个“诗意”的世界，同时也是一个“如画”的世界。

在流芳园一期建成之后，汉庭顿邀请了在美国的知名文化人为流芳园的景致进行题咏。目前在园内的多方题咏各有妙境，它们可以帮助游人领悟景点的意境以及中国园林艺术的内涵。这些匾额楹联也为流芳园增添了许多文化气息。在流芳园的设计过程中，我一边设计一边思考景点的立意和题咏诗句，以此点出景点的意境。在此，根据已经建成部分的一些景致，采用图片与诗文相配的形式来诠释流芳园的“诗情画意”。比如，反映流芳园四季美景的诗句：“春绿锁桥榭，夏翠落名园。秋色荡明湖，冬影照粉墙。”

The poetic and picturesque expressions of the garden are not only in scenic forms, but also in inscriptions on horizontal boards and couplets, etc. The inscriptions depict something beyond the garden scenes or imageries. The ancient Chinese scholars regarded the poetry, calligraphy and painting as the “scholar’s three perfections”. The practice of writing poems or inscriptions on paintings started in Ming dynasty. This particular approach integrated painting, calligraphy and poetry into a unique art form by which without stepping out of the house, the artist would be able to express his sentiments on the paper freely, the landscape on the painting would be fully presented to his eyes, and the artistic conceptions would be fully realized in his heart. With the same regard, composing the inscriptions and naming the garden scenery during Ming and Qing dynasties were an inseparable part of the garden design and construction. This approach has become an artistic way to enhance the aesthetics of the garden scenery as well as a playful activity in the process of making a garden. In fact, naming and describing the scenery of a garden will help, first, to

秋亭嘉树图　元　倪瓒　故宫博物院藏
Ni Zan (1301-1374), Autumn Pavilion sits in the fine trees, The Palace Museum

depict the artistic conceptions behind the scenery in the garden; secondly, to decorate some of the garden features such as pavilions or rockery; and thirdly, to inspire the visitor to re-create one's own new conceptions of the scenery. If you create a garden based on the poetic conceptions, while writing poetry according to the garden scenery then the poetry becomes a imaginary garden, the garden a visible poem. The Chinese garden is not only a poetic world, but also a picturesque world.

After the completion of the phase one of Liu Fang Yuan project, the Huntington had invited some well known Chinese scholars in the United States to compose and write inscriptions and names for the garden structures and scenery. Most of the wonderful inscriptions and couplets in the garden will help visitors to understand and appreciate the artistic conceptions of the garden and the Chinese garden culture as well. These inscription and couplets have enhanced the literary and scholar atmosphere in Liu Fang Yuan. During the design process, I worked on the design and, at the same time, the poetic compositions for most of the scenes in the garden. Here in this book of *Liu Fang Yuan*, I have selected some of the photographs showing the completed sections and garden features of Liu Fang Yuan today, and composed the poetic lines for each of these pictures to present them in this chapter. For example, regarding the four pictures which present the scenery in each season, the poetic verses read like: "Spring greens lock the bridge and pavilion; summer emerald greens mantle the famous garden; autumn foliage colors ripple on the clear lake; and winter shadows reflect on the white-washed walls."

流芳园 – 曲桥与石拱桥
LFY - The zigzag Bridge and the Arched Stone Bridge

水天映秀园，和静藏幽深 The elegant garden shines upon the water and sky, the harmony and restfulness concealed in the deep and secluded.

流芳园 – 湖光山色与园林融为一体

LFY - View of the mountain and waters blends with the scenery of the Garden

斜桥跨曲水，祥云照芳湖

Zigzag Bridge across the winding water; auspicious clouds shine on the fragrant lake.

流芳园 – 曲桥芳湖、亭台楼阁

Liu Fang Yuan - Bridges, the lake and pavilions

蕉荫晚翠，粉墙深院

Plantain shades the evening greens
behind the white washed walls are the deep courtyards.

流芳园 - 粉墙、黛瓦、芭蕉叶

LFY - the white-washed-wall, the black roof tile and the plantain leaves

临水最相宜，暮色越台空 Best fit is the waterside, in the Yue Tai no one is in sight.

流芳园 – 清越台

LFY - the Qing Yue Tai

绮窗影相照，梁上花吐芳

The lattice windows cast shadows on each other; the flowers on the beams diffuse fragrance.

流芳园 － 玉茗堂内景

LFY - the interior of the Tea House

雕梁花罩夺天工

Wonderful workmanship of the carved beams and screens excel nature.

流芳园 － 清越台木雕

LFY - the wood carvings on Qing Yue Tai pavilion

春绿锁桥榭

Spring verdant greens lock the bridge and pavilion.

夏翠落名园

Summer emerald greens mantle the famous garden.

秋色荡明湖

Autumn foliage colors ripple on the clear lake.

冬影照粉墙

Winter shadows reflect on the white-washed walls.

长龙卧碧，花窗漏影

The long dragon crouching in the bluish-green,
lattice windows leaking the garden scene.

云辟竹影，白石苍古

Wall of Clouds with bamboo shadows, and the aged gray rock

苍松画舫，芳草池塘

Deep green pines behind the boat-shaped pavilion,
the pond is flanked by grass of perfuming and pliant.

碧水云天上，曲桥叠孔桥

The bluish water mirrors the clouds in the sky,
the zigzag bridge overlapped by the arched bridge.

日落庭院静，信步曲廊深

After the sunset the courtyards are quiescent,
with one walking in the crooked corridor toward the deep end.

浓绿相会，芭蕉几许

Dense greens assemble a few plantain trees.

柳绿桥阁间，鱼乐芳湖中

Willow green is between the bridge and pavilion,
fish are happy in the fragrant lake.

鱼乐桥头，无尽烟波

Bridge of the Joy of Fish winds to the endless mist.

鱼乐、人闲、高阁静

Here are happy fish, the relaxed people, and the placid tall pavilion.

春水绿波花影外，藻舟待渡运逍遥

Beyond the peach blossoms ripples on vernal lake spreading,
the boat-shaped pavilion is ready to depart to carry leisurely carefree people.

日长风静，莲花闲相望

The day is spreading its splendor, the air stirs no longer,
the lotus flowers glimpse at each other.

天上有行云，水底看山影

In the sky float the clouds,
in the mirror underwater are mountains, ridge upon ridge.

秋已暮，红稀香少

In the late autumn, there are sparse reds and a little fragrance.

古木苍烟，残荷藏幽

Aged trees in the gray mist, weathered lotus conveys the quiescent.

雕镂映斜阳

Carved gatehouse sits in the oblique twilight.

庭院日上头，满地阑干影

In the sun-bathed courtyard are strips of shadow in the railed corridor.

飞瀑出岫

Waterfall gushing from the mountain peak.

水流云在

Tumbling waters merge into the upper clouds.

飞檐展翅

Eaves on the wing.

玉洞含秀

The moon gate frames the garden beauties.

茅亭涤虑

The thatched pavilion washes away worries.

重阁倚石

The double roofed pavilion reclines against a rock.

楼船待渡

The boat-shaped pavilion is ready to cross the water.

清音环中

The tuneful music comes out of the ring-shaped moon gates.

湖石铺天

The Taihu rock is embedded in the sky.

千年之韵

The rhythmic vitality goes on through thousands years.

当檐交相应

The eaves respond to one another.

秀宇穿晴空

The beautify roof protruds the clear sky

“流芳园”门额

Brick carved the name of "Liu Fang Yuan"

云墙细部

The detial of the Wall of Clouds

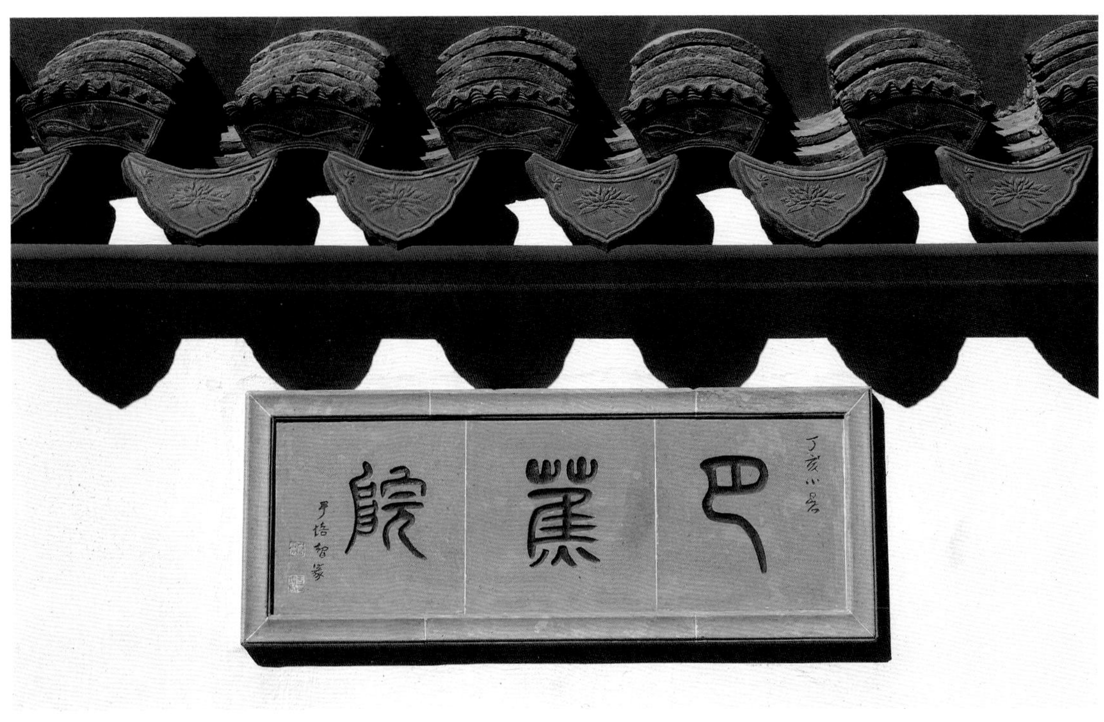

“芭蕉院”门额

Brick carved name of "Plantain Court"

园墙细部

Detail of the Garden Wall

“玉镜台”砖雕

Brick carved name of "yu jing tai"

(Terrace of the Jade Mirror)

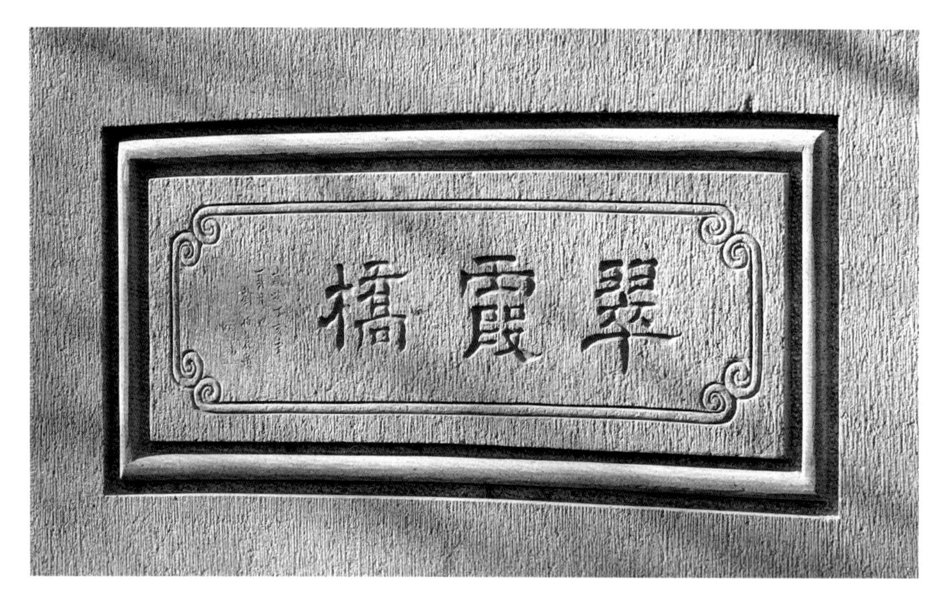

“翠霞桥”石刻

Stone carved name of "cui xia qiao"

(Bridge of Verdant Mist)

海棠花图案铺地

Paving pattern of begonia flowers

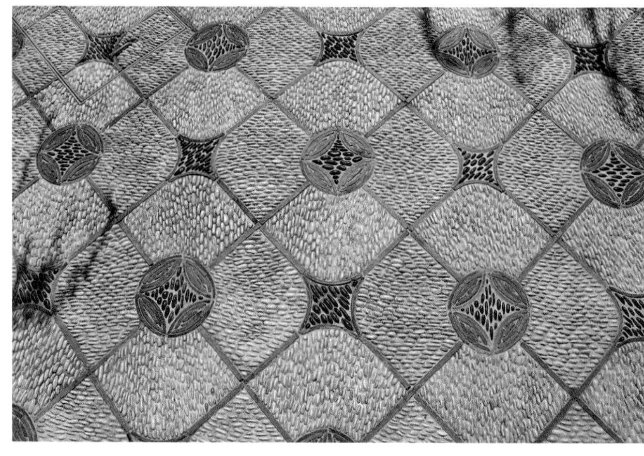

铜钱图案铺地

Paving pattern of old coins

点石小品
Garden detail

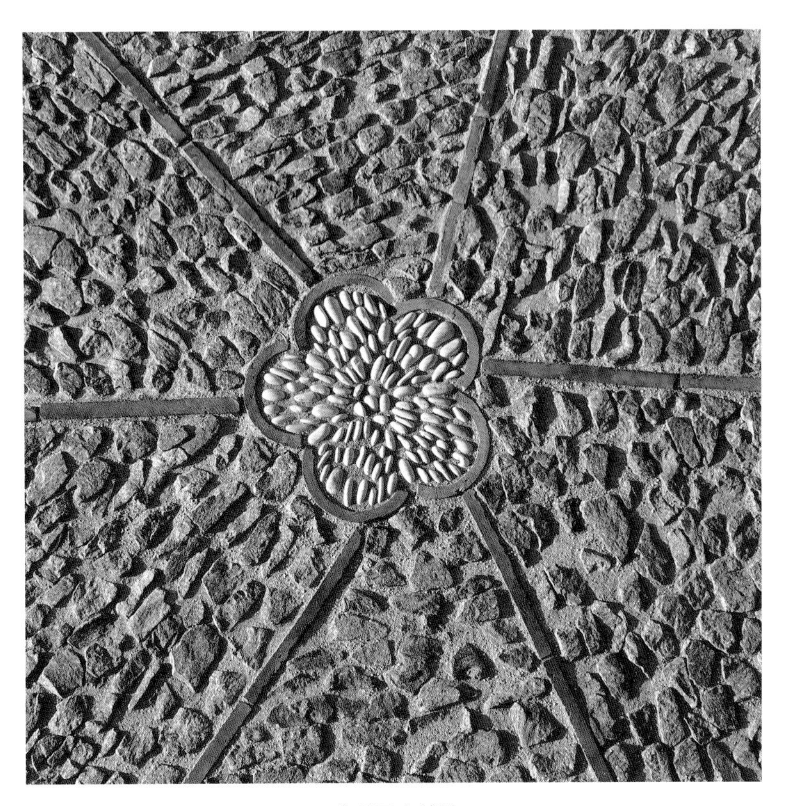

梅花图案铺地
Paving pattern of plum flower

流芳园中景点匾额

The inscribed boards of names for the pavilions and corridors

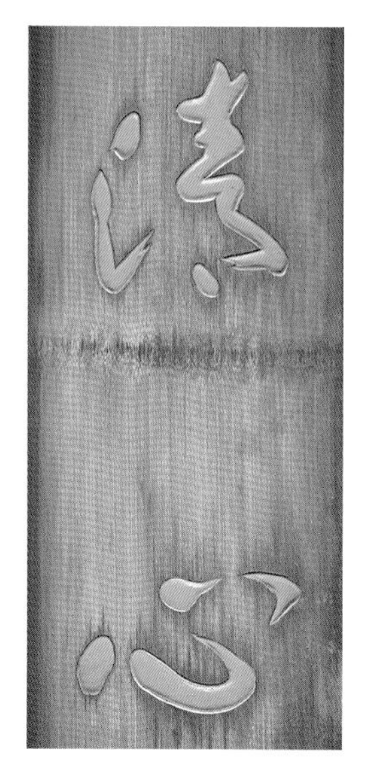

楹联细部

Details of couplets

"爱莲榭"木雕屏风
Wood carving screen inside the Love for the Lotus Pavilion

苏州名园主题木雕细部
Thematic wood carvings on Suzhou gardens

木雕菊花
Wood carved Chrysanthemum

第七篇

流芳园规划设计部分图录

Chapter Seven

Master Planning and Design Drawings of the Garden

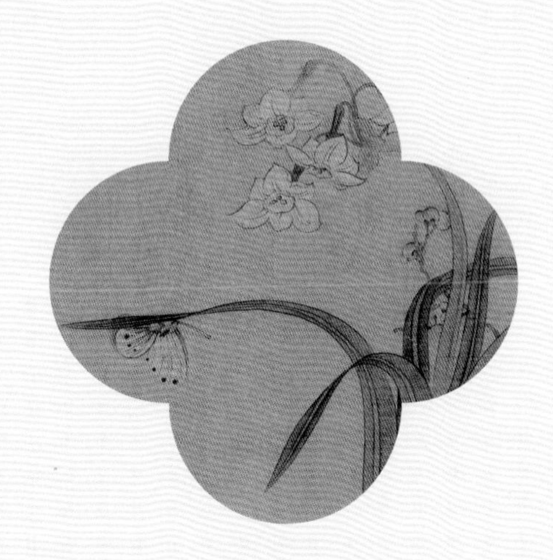

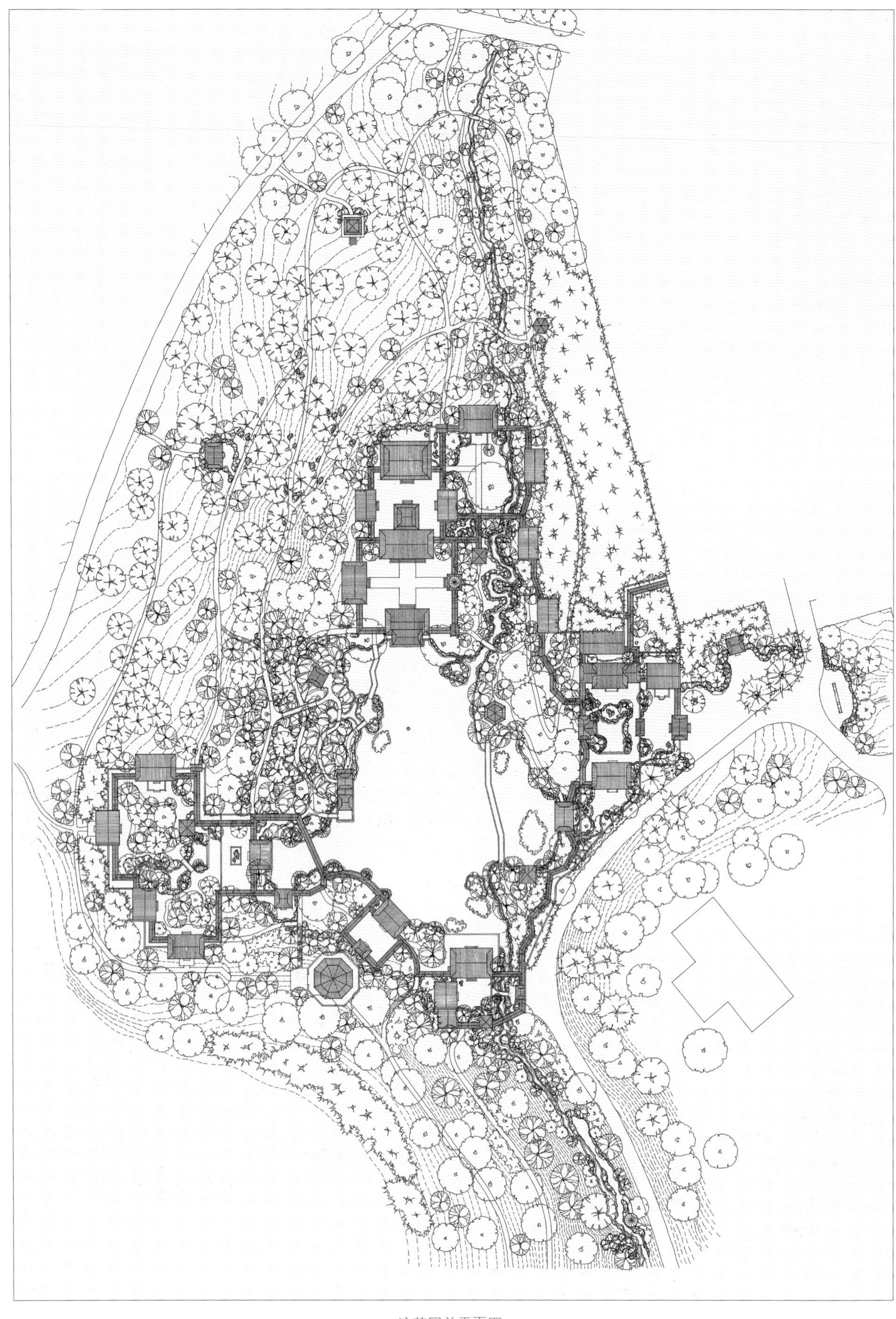

流芳园总平面图

The Overall Layout Plan of Liu Fang Yuan

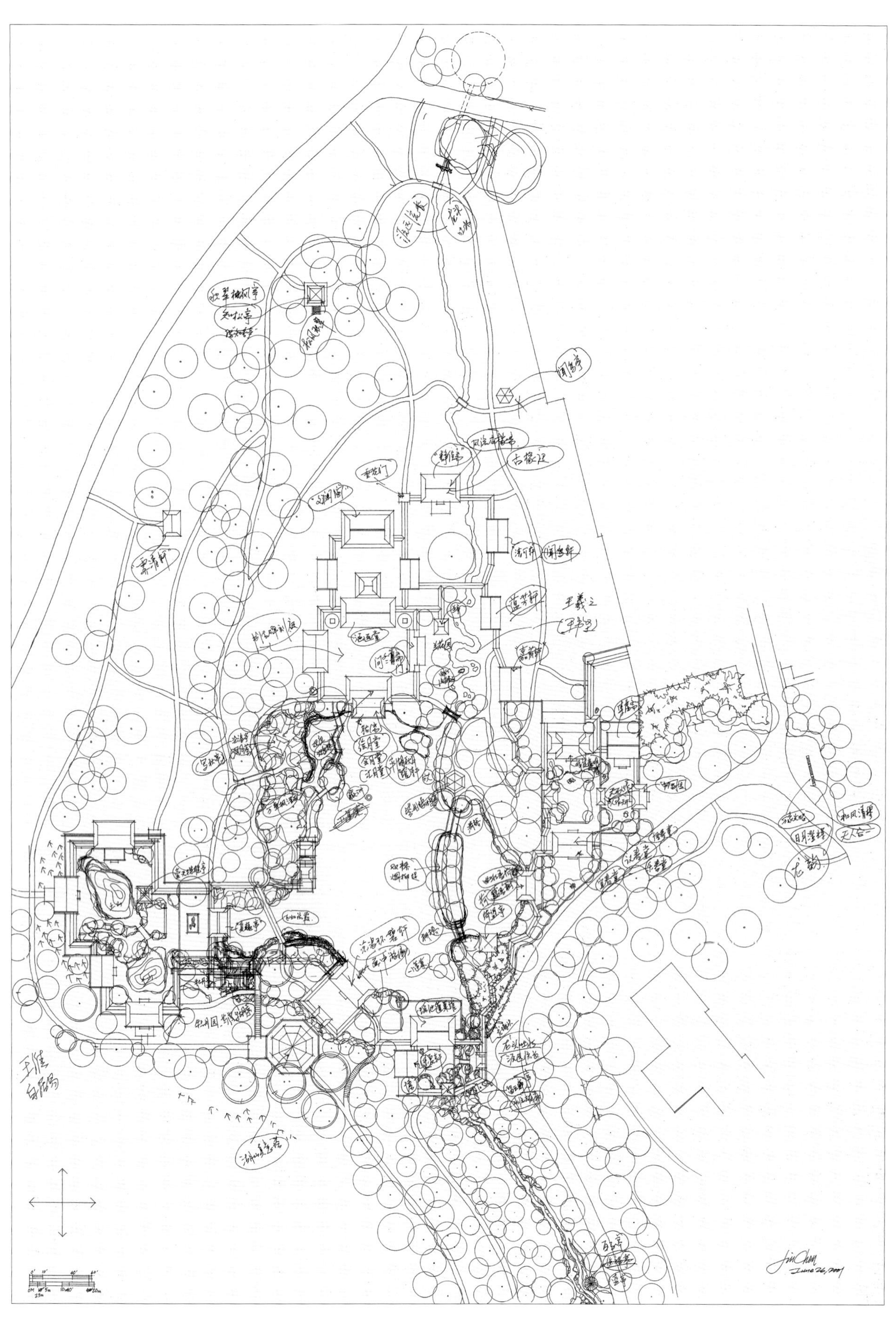

流芳园景点题名构思图

The Naming Plan of Scenes in Liu Fang Yuan

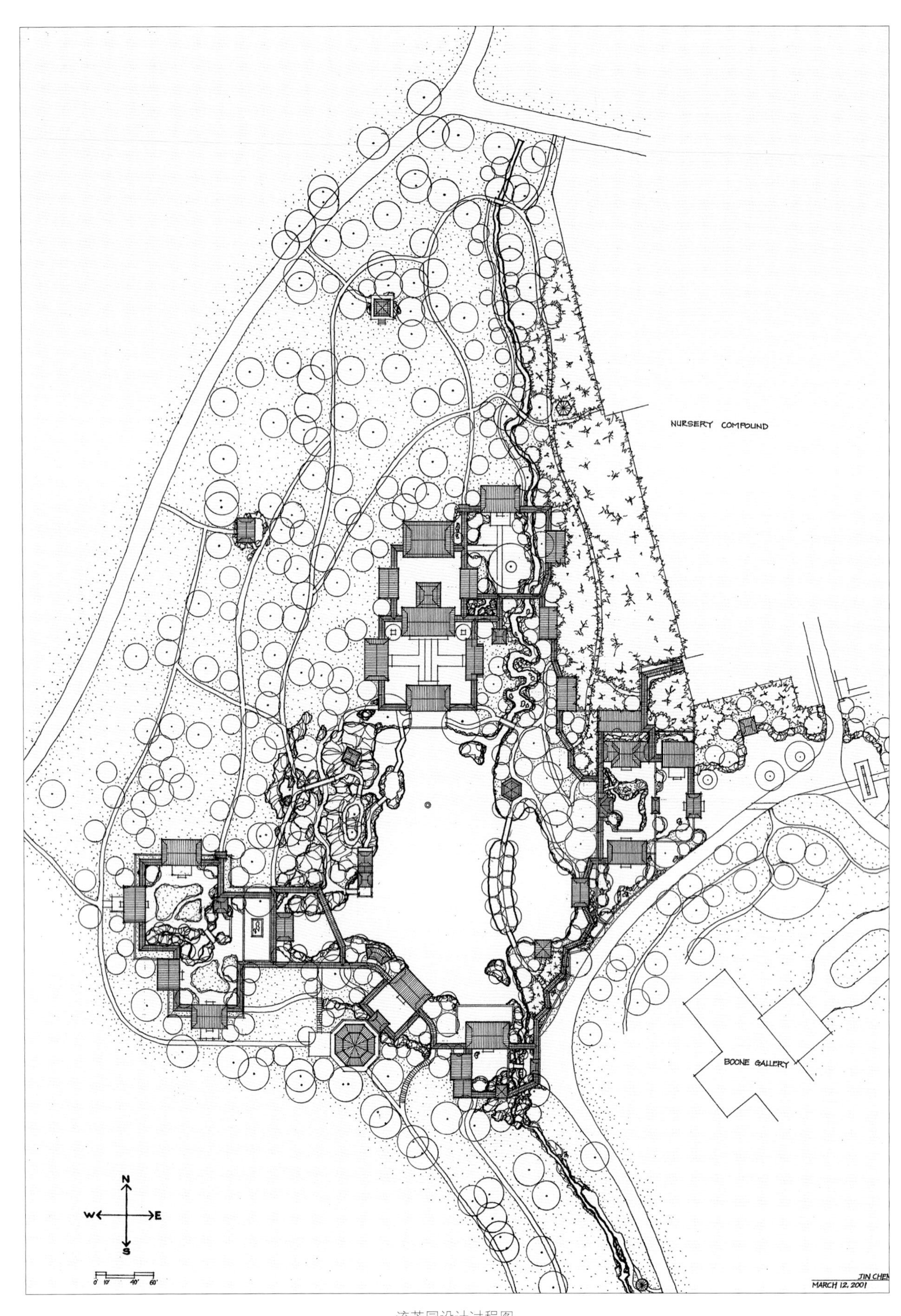

流芳园设计过程图

A later version of the Layout Plan of Liu Fang Yuan

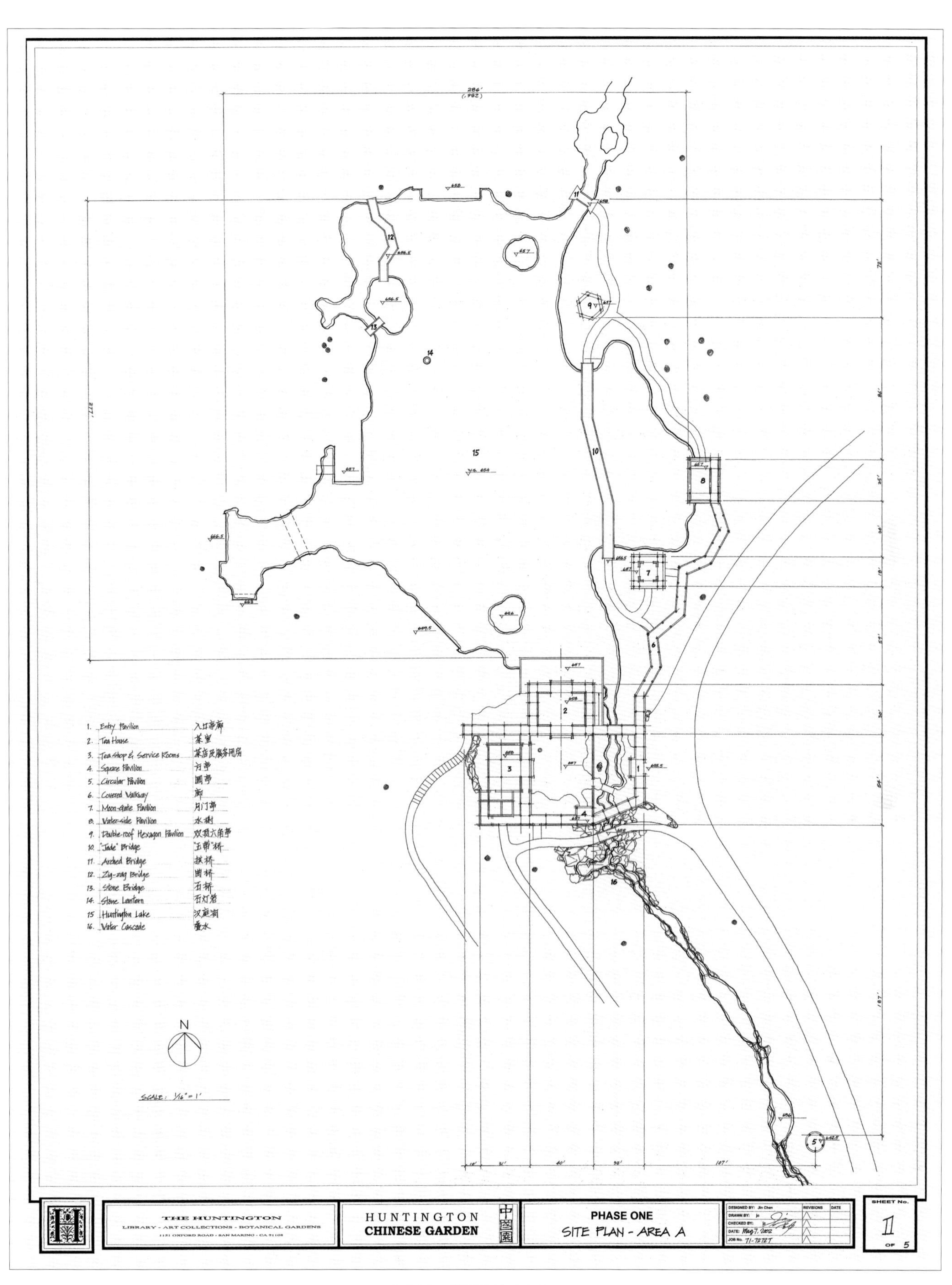

流芳园一期平面布置图之一

Liu Fang Yuan - Phase One plan 1

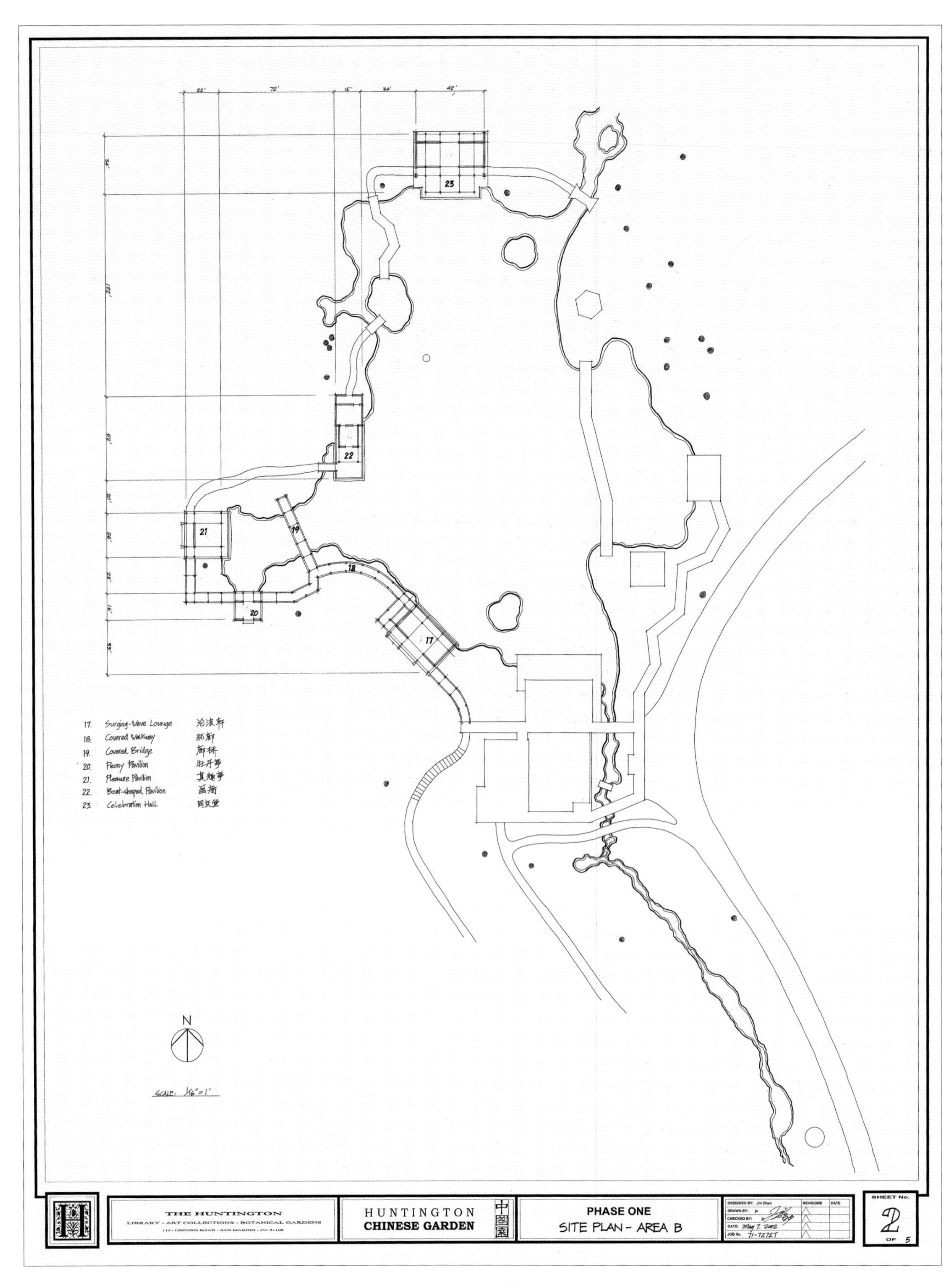

流芳园一期平面布置图之二

Liu Fang Yuan - Phase One plan 2

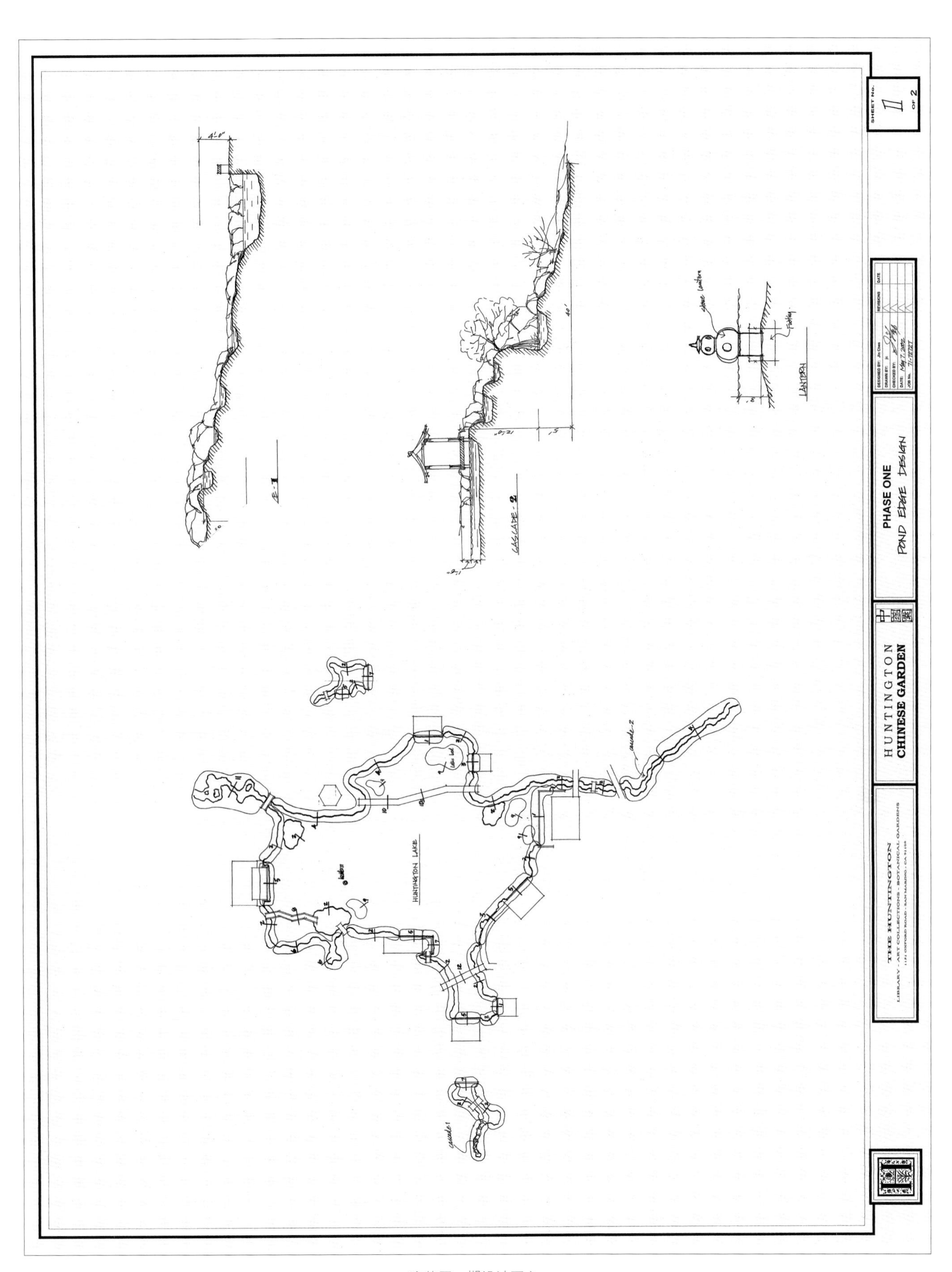

流芳园一期设计图之一

Liu Fang Yuan - Phase One design drawing 1

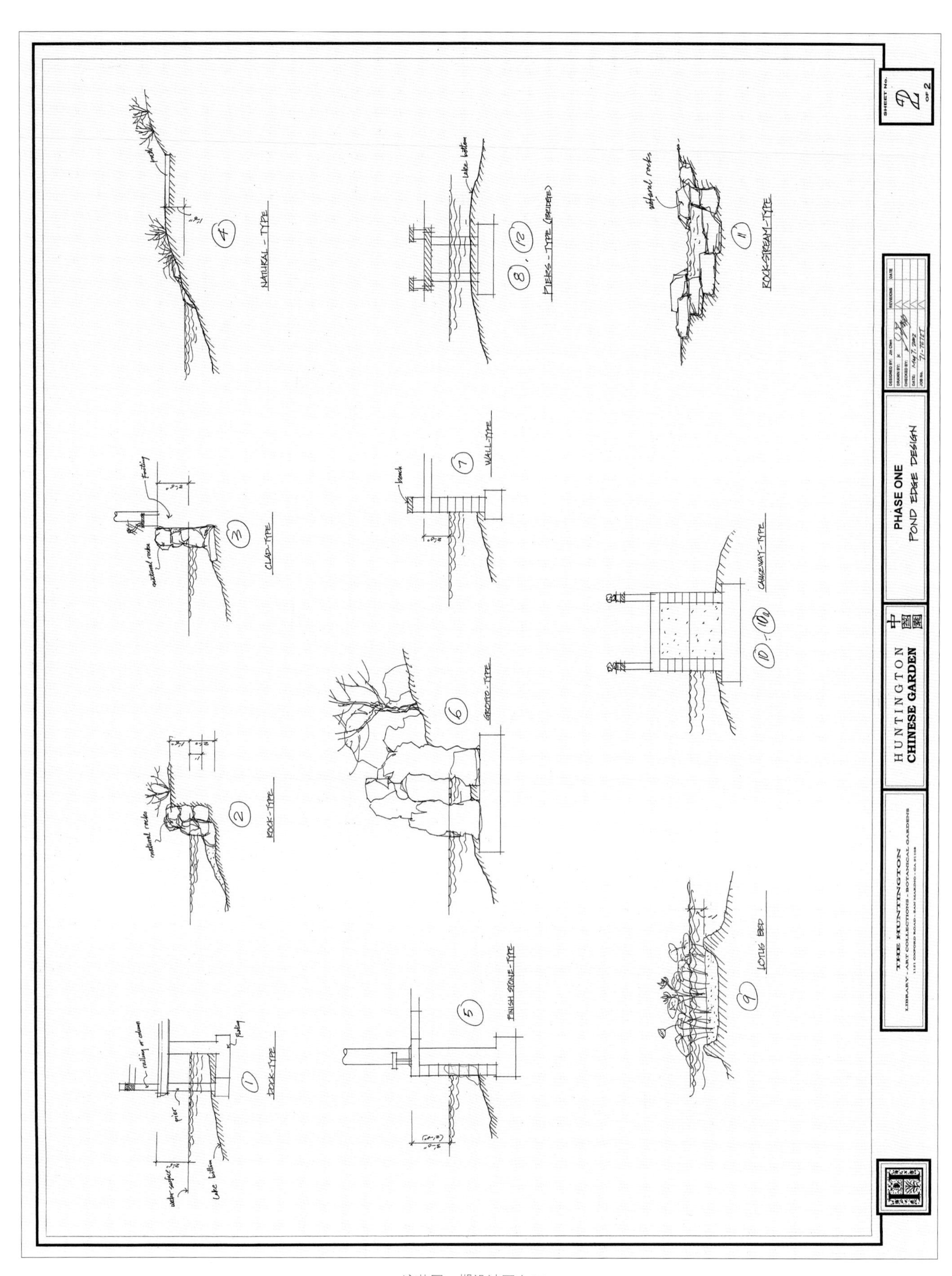

流芳园一期设计图之二

Liu Fang Yuan - Phase One design drawing 2

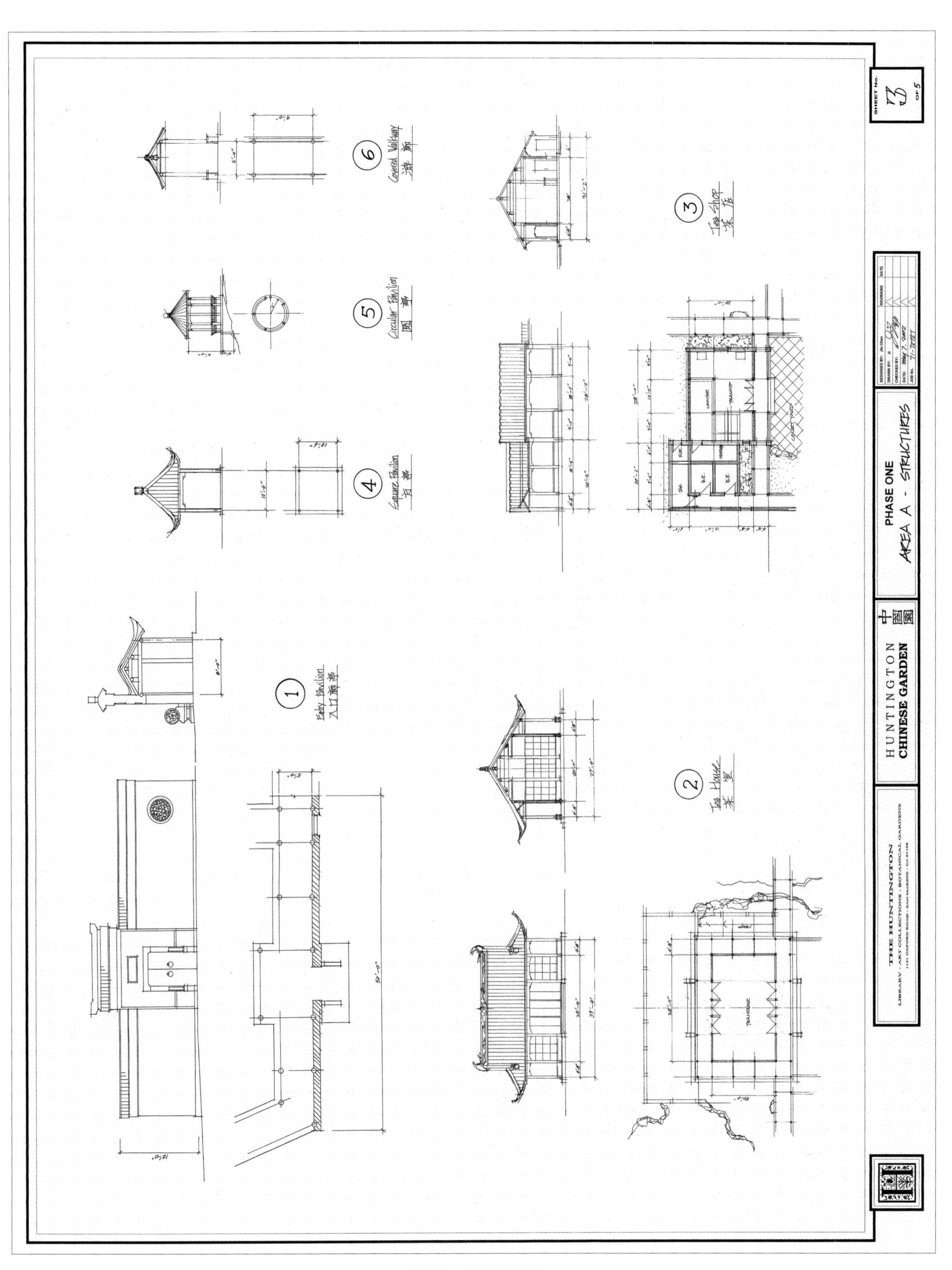

流芳园一期设计图之三

Liu Fang Yuan - Phase One design drawing 3

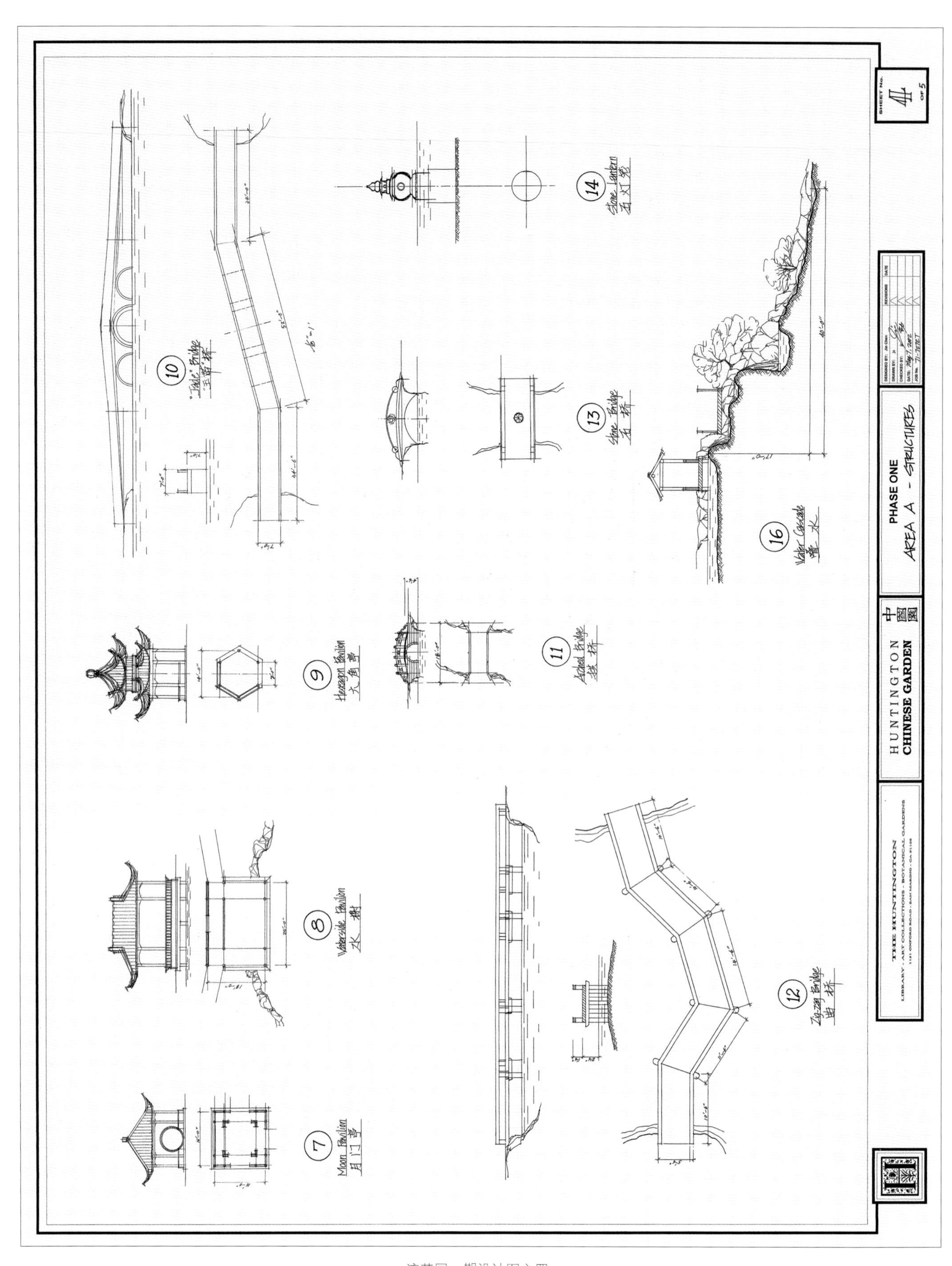

流芳园一期设计图之四

Liu Fang Yuan - Phase One design drawing 4

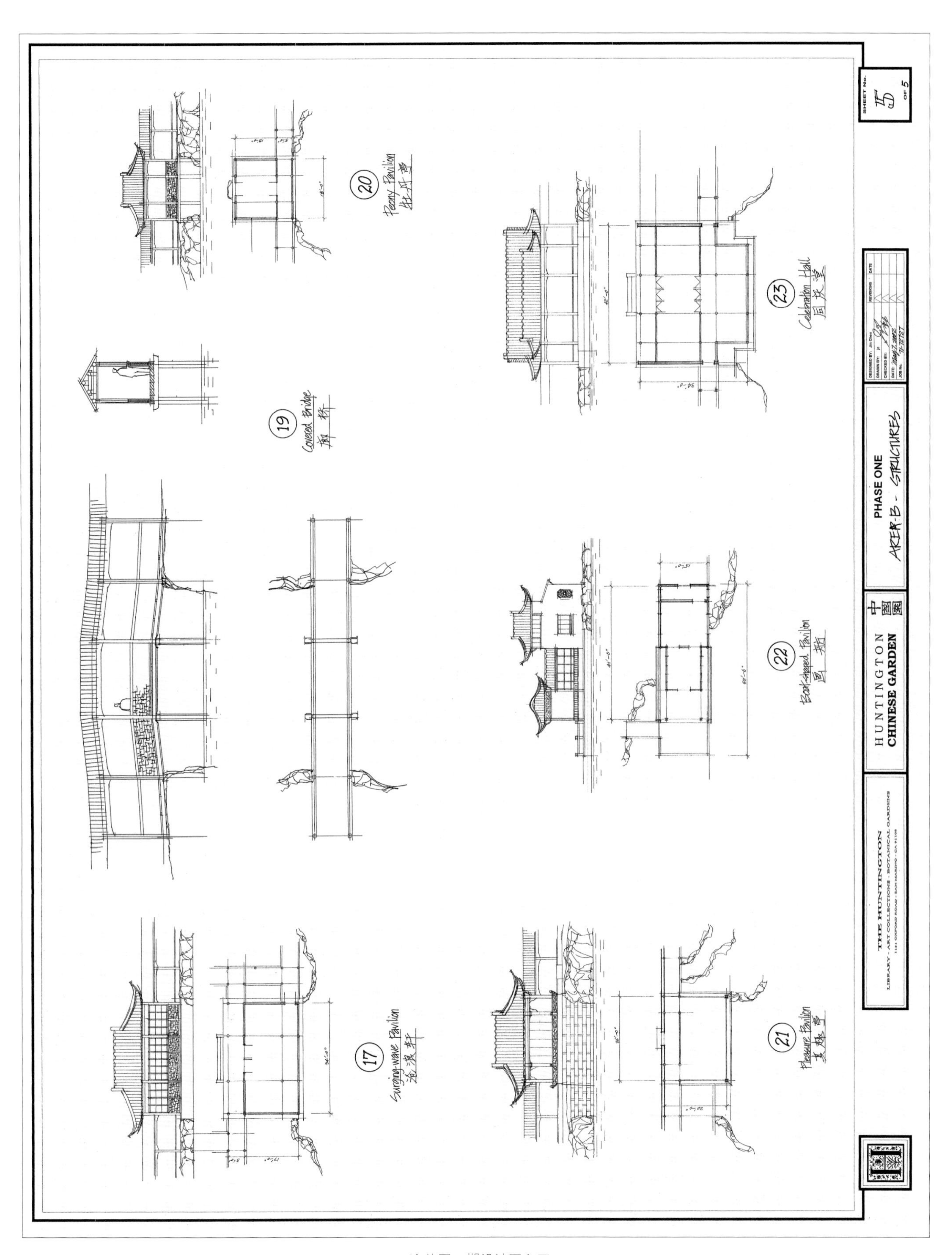

流芳园一期设计图之五

Liu Fang Yuan - Phase One design drawing 5

流芳园建筑设计图

Building Design Drawings For Liu Fang Yuan

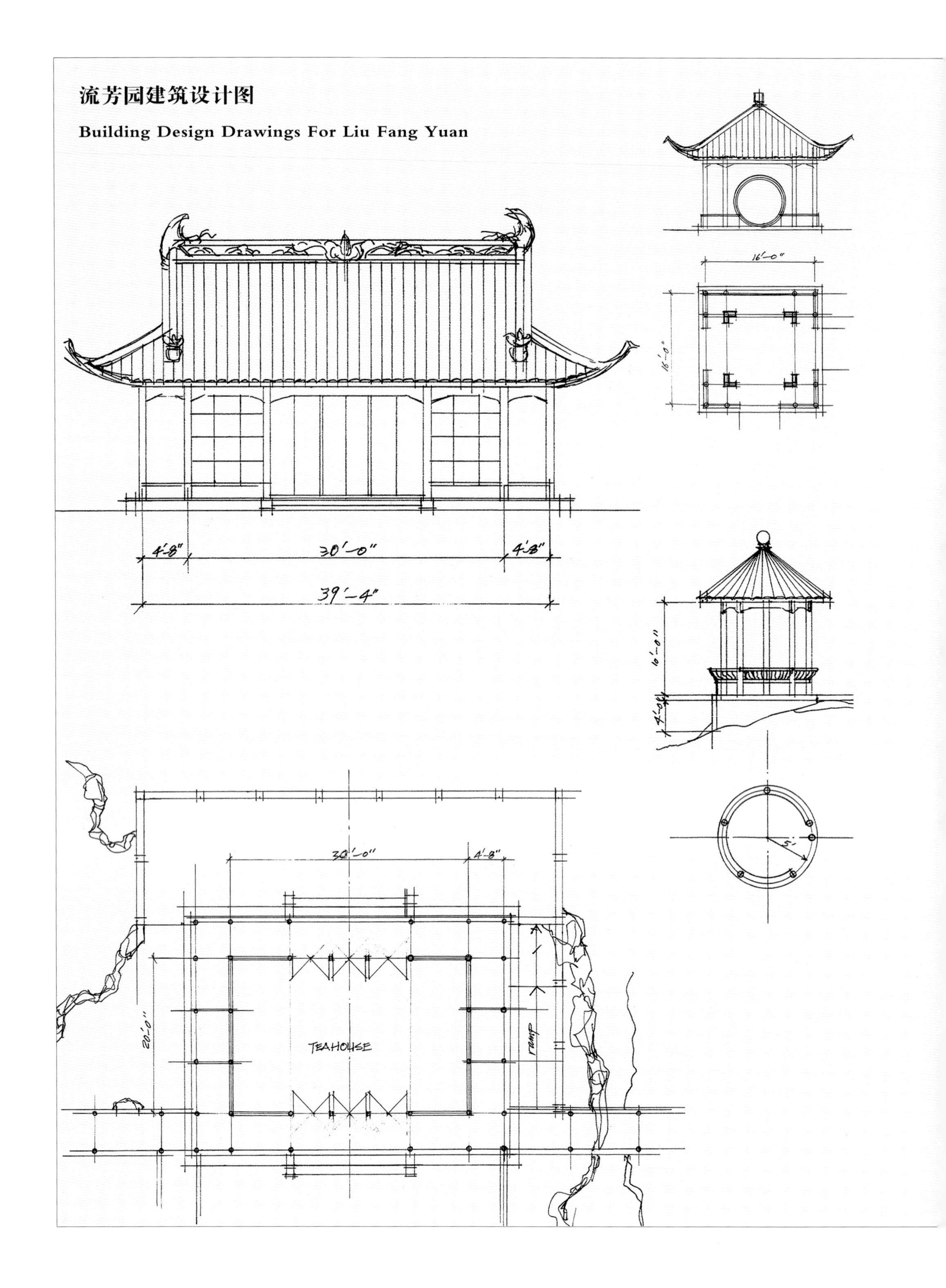

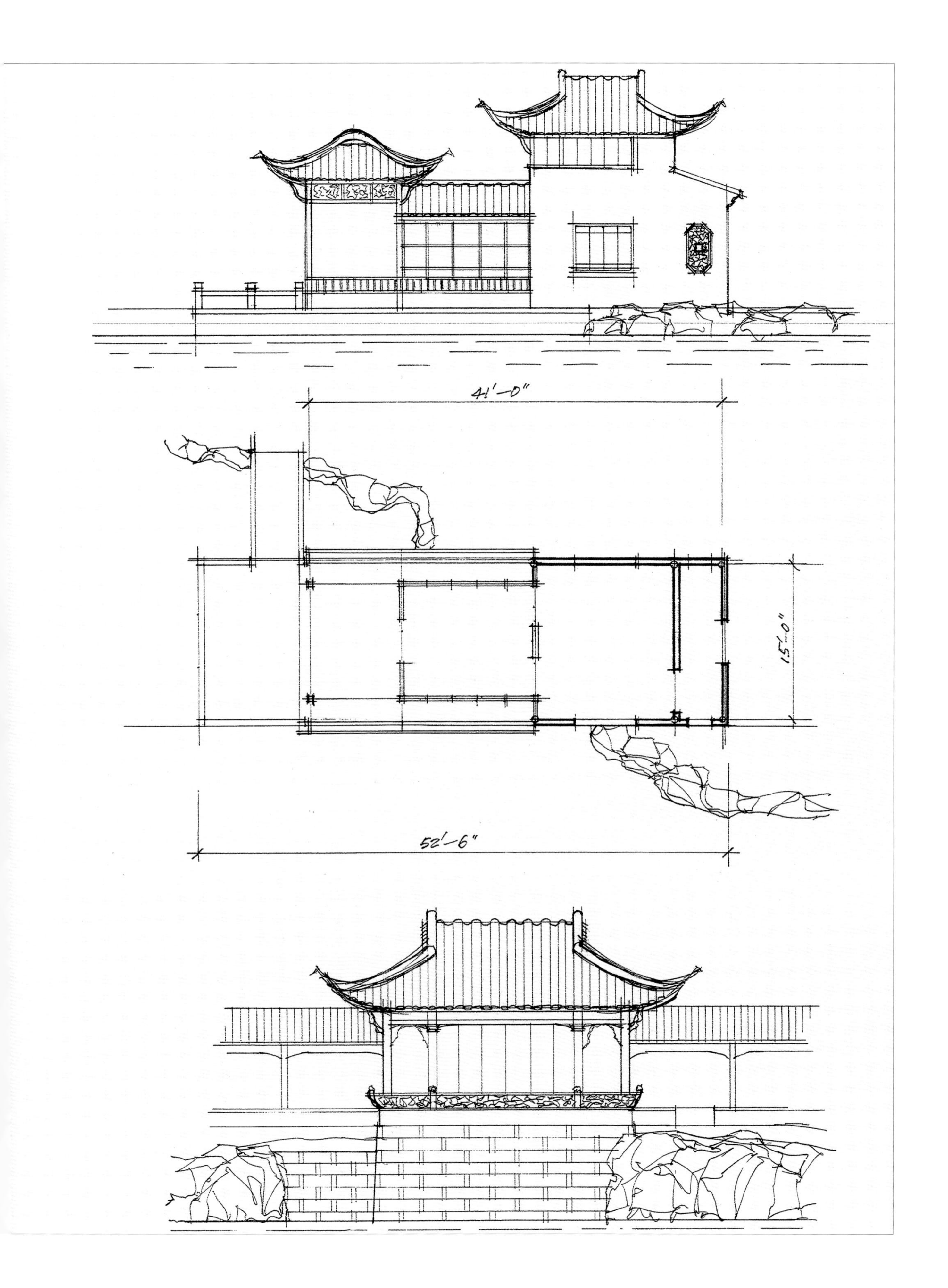

41'–0"
15'–0"
52'–6"

流芳园景点构思图“曲水香荷”
LFY - The scenery of "winding waters hold the fragrant scent of lotus"

流芳园景点构思图“水流云在”
LFY - The scenery of "tumbling water against upper clouds"

流芳园景点构思图“玉带接秀”
LFY - The scenery of "jade-ribbon-bridge linking beauties"

流芳园景点构思图“瑶池蓬莱”
LFY - The scenery of "the Yao Chi wonderland"

香榭对横塘，静林望高云

The fragrant pavilion stands over an expanse of pond, the static forest look up to the high clouds.

流芳园 – 爱莲榭
LFY - the Love for the Lotus Pavilion

后

记

Epilogue

缘分
THE PREDESTINATION

2000年2月，为了给美国俄勒冈州波特兰市的中国园林（“兰苏园”）选择中国品种的植物，我和波特兰的两位植物专家翔·侯庚先生和帕克先生一起到加利福尼亚州南部进行考察。我们走访了数家苗圃之后，侯庚先生建议我们一起去拜访汉庭顿植物园主任詹姆斯·富尔森先生并考察一下汉庭顿。之前我只听说过汉庭顿这样一个文化艺术机构，这次终于有机会见到其真面目了。当我们的车驶入那标识有一个大“H”的门楼，沿着迷宫似的

In February 2000, in order to find some planting materials for the Portland Chinese Garden – Lan Su Yuan or the "Garden of Awakening Orchid", I had traveled with two Portland botanists Mr. Sean Hogan and Mr. Parker to Southern California and visited a few nurseries in Los Angeles area. One day, Mr. Hogan suggested we visit the Huntington Library and meet with Botanical Gardens Director James Folsom to see if we could find the plants we sought for there. I had heard about the Huntington Library but never visited this place until this time. When our car drove through the main gates with the huge letter "H" plaque and approached the botanical compound through labyrinthine driveways, I was incredibly impressed with the lush plant selection on both sides of the road. We went directly to the nursery to meet with James Folsom, who turned out to be a quick-

流芳园 - 入口前庭

Liu Fang Yuan - the front yard of the Entrance

车道行驶的时候，路两边丰富的植物开始引起了我的极大兴趣。当时我们并没有去汉庭顿的主楼图书馆和艺术馆，而是径直到了植物园的办公室和苗圃，在那里我们见到了富尔森先生。他是一位看上去总是激情四射、思维敏捷、侃侃而谈的学者。我们一直用英文直接交谈，不过他提到的部分植物名称我并不熟悉，侯庚和帕克先生及时帮助解释了部分植物品种。当富尔森先生得知我正在波特兰负责兰苏园项目后，他兴致勃勃地介绍说汉庭顿也在筹划建造一座中国园林，并跟我进行了深入的交谈。之后，他说下午他们计划请当地一家电视台的记者和摄制组来汉庭顿拍摄介绍汉庭顿植物园的片子。这时，富尔森先生产生了一个奇特的想法，即请电视台对我们进行一次采访，让我和侯庚先生介绍一下波特兰中国园的项目。我们欣然答应了他的提议并在采访中简要地介绍了“兰苏园”的实施情况。采访之后，富尔森先生直接问我是否有兴趣来汉庭顿工作并负责中国园的项目。我思考了一下，便表达了我对汉庭顿中国园项目十分感兴趣，并愿意发挥自己在中国园林艺术方面的特长，为该项目做出贡献。富尔森先生随即表示了他将开始具体安排相关的事宜。第二天，我们返回到了波特兰市。

七个月之后，即 2000 年 9 月，波特兰的“兰苏园”正式建成对公众开放。开放仪式当天，中国驻美国大使李肇星先生，苏州市长陈德铭先生，波特兰市长卡兹女士以及其他贵

witted and eloquent scholar that almost always looked enthusiastic. We talked to each other directly in English, but when I had some difficulties recognizing a few botanical names in the language, Mr. Hogan and Mr. Parker came to my rescue with explanations. When Mr. Folsom learned that I was working on the Portland Chinese Garden for the City of Portland as the project coordinator and design consultant, he told me that the Huntington was planning to build a Chinese garden too. Then Mr. Folsom asked me questions about the Portland project as well as ones about Chinese gardens in general. Sometime later, he told us that there was a crew from a local TV station coming to the Huntington to take pictures and interview some of the botanical staff for an educational program to be shown on the station. At this point, Mr. Folsom had the idea of asking the TV crew to interview and videotape us. He asked for Mr. Hogan and me to talk about the Portland Chinese Garden project. After the interview with the TV crew, Mr. Folsom asked me if I was interested in coming to the Huntington to work on the Huntington's Chinese Garden project. After a thought, I told Mr. Folsom that I would be happy to come and work on the project and apply my design experiences and knowledge of Chinese gardens to make a contribution to it. Mr. Folsom told me that he would arrange this matter soon. Next day, we returned to Portland.

Seven months later, in September 2000, the Portland Chinese Garden or Lan Su Yuan was completed and opened to the public. During the opening ceremony, Mr. Li Zhaoxing of the China's ambassador, Mr. Chen Deming the Mayor of Suzhou, Ms. Vera Katz the Mayor of Portland and

宾都来参加了开园仪式。至此我在兰苏园项目上五年的工作也正式结束。所以，我于 2000 年 10 月移居到加州洛杉矶的圣马利诺市并开始了在汉庭顿的工作。在随后的三年时间里，我致力于中国园项目——流芳园的总体规划设计以及相关的各项工作当中。

我有幸参与汉庭顿流芳园项目确实是一种缘分和巧遇。讲到“缘分”，首先是因为我在大学本科学的是园林设计，我于 1983 年毕业于同济大学建筑系园林专业。在中国传统园林学习方面，有幸得到同济大学教授、中国古典园林一代大师陈从周先生的亲自传授和指导。特别是在 1982 年做上海古典园林豫园的东部修复设计课程中，陈先生亲自到豫园现场进行授课和指导，使我通过实例的设计对中国古典园林的造园思想和手法有了深刻的理解和一定的掌握。其次，我自 1989 年到美国之后相继获得了麻省大学的景观设计硕士和俄勒冈州立大学的建筑学硕士两个学位，并在美国的景观和建筑设计公司工作数年，获得了美国的相关规划设计工作经验。尤其是在 1995 年至 2000 年期间，我在美国俄勒冈州波特兰市政府负责波特兰中国园——“兰苏园”的项目。作为设计顾问和项目经理，我与苏州园林公司以及美方团队的设计师们和施工人员一起，从设计的落地、与场地的结合、建材的选择、植物的选择和配置，到将中国传统园林的设计和施工与美国当地的建筑及抗震规范相结合等方方面面，一步步完成了各项挑战，终将一座颇具苏州园林风格的中国园林建成并呈现

other distinguished guests attended the ceremony. By then, I had completed my five-year task on Lan Su Yuan project. In October 2000, I moved to San Marino, Los Angeles and started to work on the Huntington's Chinese Garden project. In the following three years, I focused on the master planning and design of Liu Fang Yuan. In addition, I also participated in some of the related activities such as presenting the design and the project to the Huntington staff, local communities and fundraising events.

I was fortunate to able to join the Huntington team to work on Liu Fang Yuan project. I felt like this was predestined and a chance encounter. Speaking of "predestination", first, my major in my undergraduate years was to study garden design and landscape architecture in Tongji University from 1979 to 1983. During those years at Tongji, Mr. Chen Congzhou, a professor at Tongji and a preeminent Chinese garden master in modern China, taught classes on traditional Chinese gardens. I was his student. In 1982, during one particular class for a real-life design project on the restoration of the eastern side of the Yu Yuan classical garden in Shanghai Professor Chen taught the class inside the garden. I learned a lot about Chinese garden design principles as well as garden culture through this real-life design project. Secondly, since coming to the United States in 1989, I had done my graduate studies in landscape architecture and architecture from the University of Massachusetts and the University of Oregon, and finished with two master's degrees. After these,

给世人，我个人也从中获得了许多实践经验。从缘分角度讲，这些知识和经验的积累，为我进行汉庭顿流芳园的设计打下了良好的基础。

讲到“巧遇”，自然是当年有幸遇到汉庭顿的植物园主任富尔森先生，以及当时汉庭顿流芳园项目所处的阶段正是需要进行总体规划和一期工程设计这样一个机遇。同时，在汉庭顿工作的三年时间里，我的工作得到了汉庭顿流芳园项目团队和其他各部门的大力协助，特别是得到了富尔森主任和时任汉庭顿总裁史蒂夫·郭必列先生的鼎力支持。在流芳园的施工图设计过程中，我还得到了中方设计团队——苏州园林设计院贺凤春院长和谢爱华副院长所领导的设计团队和陆宏仁高级工程师领导的施工方的积极配合，以及当地建筑师鲍勃·奥芬豪斯先生和吉米·弗莱先生所带领的美方设计团队的协助。可以说，流芳园项目是尽得天时、地利及人和的一个好项目。

流芳园 － 柱头雕饰

LFY - Wood carved post joint

I worked in architectural and landscape design firms in the U. S. for many years. My experience of traditional Chinese garden design and construction was further enriched during the years from 1995 to 2000, when I was working on the Portland Chinese Garden Lan Su Yuan as the project coordinator and design consultant in the City of Portland. In the project, I worked closely with the Suzhou design and construction teams as well as the American teams on the issues such as making the design and construction fit the site and adhere to the local building codes and seismic requirements, while selecting building and planting materials, and getting the building permits, etc. After the collaborations and coordination among the teams, the garden was successfully completed and opened to the public. All my experiences and knowledge accumulated to that point, therefore, had built a solid foundation for me to work on the design of the Huntington's Chinese Garden Liu Fang Yuan.

As speaking of the “chance encounter”, I think, first, it was the moment when I met with Botanical Gardens Director James Folsom at the Huntington in 2000; and second, during that time, the project of the Huntington's Chinese Garden Liu Fang Yuan was ready for its master planning and first phase design. In addition, I was fortunate to be able to work with a wonderful team during those three years at the Huntington. I also had great supports from the leaders of the Huntington institution, especially from Director James Folsom and the Huntington's President Steve Koblik. During the construction design process in the first phase of the project, I had a great collaboration with the Chinese design team from the Suzhou Garden Company, led by General Manager He Fengchun Deputy Manager Xie Aihua and senior engineer Lu Hongren; and with the American design team led by architect Bob Ray Offenhausor and James Fry. It can be said that Liu Fang Yuan is a wonderful project achieved from the best of time and opportunity, the best of site and environment, and the best of the working teams and project supporters.

感悟
The Reflections

此时此刻，我首先思考的是：为何在当今欧美的国家中，时有兴建新的中国传统式园林之举？它的意义何在？纵观中国园林文化和艺术的内涵，我认为中国园林对于当今的人居环境以及世界其他园林文化具有以下几方面的启示和意义：其一，中国园林崇尚自然和再现自然；其二，中国园林体现生态环境的可持续性；其三，中国园林强调人与自然的和谐；其四，中国园林可以提升人的心性修养和精神境界；其五，中国园林蕴含着“以人为本”的精神；最后，中国园林是中国文化博大精深的综合体现。就与世界其他文化进行交流而言，中国园林是最好的载体之一，因为它可以向人们提供身临其境地体验和多方位地认知中国文化艺术的内涵和魅力。

At this moment, I have started to reflect on why there are many traditional style Chinese gardens have been built in European countries and the United States during recent years. What is the significance of these projects? I believe that the essence of Chinese garden has various inspirations and significances to the living environment in modern times, as well as to garden cultures around the world. The Chinese garden advocates and re-presents nature; it is a sustainable environment; it emphasizes the harmony between human beings and nature; it enhances the mental as well as spiritual qualities of mankind; and it contains the spirit of a people-oriented approach. Lastly, the Chinese garden is an assemblage and comprehensive reflection of Chinese arts and profound culture. In terms of exchange and communication with other different cultures around the world, the Chinese garden is one of the best vehicles for representing the various aspects of Chinese culture and arts through the real experiences in the garden environment.

Li Qingzhao, a preeminent female poet in Song dynasty, wrote the lines: “We are intimate with the reflections in the water and colors of the mountains; our words are endless and for their immeasurable virtues.” When a person travels the world with poetic emotion, the person will integrate with all things: flowers, trees, mountains and waters will all become one’s friends. The Chinese garden represents a world where people can leisurely enjoy sightseeing, stroll free and unfettered, and like a bird in the Chinese philosophical text, the *Zhuang Zi*, “splash for 3,000

流芳园 - 屋脊细部
LFY - Roof ridge details

宋代著名女词人李清照在《怨王孙》一词中曰："水光山色与人亲，说不尽，无穷好。"当人们以一种诗意的情怀去顺天游的时候，人与万物相融相即，天地间的一草一木、一山一水都能成为人的朋友。中国园林就是这样一个世界，让人畅游，像庄子一样逍遥地游，"水击三千里，抟扶摇而上者九万里"；中国园林就是这样一片天地，让人静观，像陶渊明一样自在地观，"采菊东篱下，悠然见南山"；中国园林就是这样一个家园，让人顿悟；中国园林就有这样一种境界，让人忘我；中国园林就有这样一种意境，让人畅神。中国园林讲求幽深清远、宁静天和。人在园林中，自有妙悟，中得心源。

中国园林是反映中国人的哲学思想、艺术情趣和文化生活的一片真实的天地。读诗文、赏绘画、听戏曲、游山水、品奇石、观花木，这些都是中国园林文化的组成部分。"读懂"

miles and soar 9,000 miles high." It is a place where one can contemplate in solitude, like poet Tao Yuanming who leisurely looked at the Southern Mountains, "while picking chrysanthemum by the eastern fence, my gaze upon the Southern Mountains rests." The Chinese garden is a homestead where people's consciences can be awakened to contemplate. It provides a realm that makes people selfless. It creates artistic conceptions that allow people to be rapt and gratified. The ensemble of the Chinese garden captures the seclusion, tranquility, the spirit of life and the natural consonance. People inside a Chinese garden will notice their own subtle comprehensions, and create something new their heart.

The Chinese philosophical thoughts, artistic interests and cultural activities are reflected in the real world of the Chinese garden. Reciting poems, appreciating paintings, listening traditional operas, strolling among hillocks and waters, admiring rare stones, and watching flowers, etc., all these activities take place inside the gardens, which are interrelated and integral part of the whole Chinese garden culture. "To understandably read" a Chinese garden, one can truly capture not

一座园林也就读懂了中国文化和艺术的精神！曲折蜿蜒的小径，斗折萦回的回廊，起伏腾挪的云墙，古朴雅致的石桥，幽深和静的庭院，神采各异的亭榭，钟灵毓秀的山水，婉转绵延的溪流，虬曲盘旋的古树等，汉庭顿流芳园都可以呈现给游人。东晋大画家顾恺之提出了著名的绘画观点："以形写神"。园林亦同此理。中国园林以山水花木、建筑庭院、假山奇石这些有形的东西去"写"出天地的自然气韵和人的真性情。园林要能表达出形之外的意境，以神统形，以意融形。朱良志先生对中国艺术的形外之神韵讲得十分透彻。他在《中国艺术的生命精神》中说："人的灵性之传达，是中国艺术的根本"，"艺术必须有内在的涵蕴，必须有特殊的寄托，必须栖息着人的心灵"。

计成在《园冶》中云："三分匠七分主人"，"多方景胜，咫尺山林，妙在得乎一人，雅从兼于半士"。园林设计师本是以图纸和图片作为传达设计思想的主要载体，然而，吾师陈从周先生说："我认为研究中国园林，似应先从中国诗文入手，则必求其本，先究其源，然后有许多问题可迎刃而解。如果就园论园，则所解不深。"造园者或设计师，即"七分主人"，若不懂中国美学，没有诗文基础，则设计灵感又自何来？本人在流芳园的设计过程中，努力将"设计师"与"文人"结合为一体，不仅以图纸表现设计，而且以"诗情画意"来构思园林意境并为园中各个景点题名，将"文思"融进了设计思路当中。每一座园林应该有其所"绘"的景和所"写"的情。中国的造园，山取奇雄崛健之势，石取瘦、漏、透、

only the outer appearances of the subjects but the inner essences of Chinese culture as well as the spirit of Chinese arts! The meandering paths, zigzag corridors, undulating cloud walls, elegant bridges, tranquil courtyards, distinct pavilions, marvelous hillocks and waters, gently stretched streams, and the gnarled oak trees, all these are assembled inside Liu Fang Yuan and presented to visitors. Gu Kaizhi (AD 345—406), a master painter of Eastern Jin dynasty, made a famous point about painting: "Forms evoke the spirit of life and vital rhythm (of the object)". It is the same for a Chinese garden. The tangible and visible things in a garden such as the hillock and waters, flowers and trees, buildings and courtyards, and rockeries, all evoke the vital rhythm of nature and the true emotions of human being. A garden should present something artistic and meaningful beyond visual pleasure from the imagery and scenery. Let the spirit of life to command the formation and the connotation integrates the forms. Zhu Liangzhi, in his book of *The Spirit of Life in Chinese Arts*, has made eloquent remarks about the spirit within art forms: "The fundamental issue in Chinese art is to convey the human's spirituality. Arts are something that must have implications, something with special sustenance, and places where the mind and soul can inhabit."

Ji Cheng said in *Yuan Ye*: "three-tenths of the work is the workmen's, seven-tenths is the master's," "the magnificent scenes on all sides; within very limited spaces, the mountains and forests are naturally presented; the wonderfulness of all these derives almost from one designer's

皱之灵气，水取蜿蜒曲折之秀，花取人格气质之意，树取苍古入画之态。这一切都在立基、理水、掇山、选石、莳花、择树之中，都在“奴役风月，左右游人”之间。

艺术是人为的。中国艺术追求由“人”而达于“天”即自然之境，通过人对自然的感悟、参与和再创造，达到与自然合而为一的境界。故中国园林艺术追求“虽由人作，宛自天开”。“天开”、“天趣”、“天巧”、“天工”之“天”之境界是园林的本源。经过对流芳园的园林设计，我对中国园林艺术和造园思想有了进一步的理解和感悟：造园，乃夺天工、乘地势、极人力之工；造园，以心源为本，造自然之相，造心中之相；造园，必使水得萦

imagination, and the other half from the owner's elegance". A garden designer of course uses drawings and graphics as his or her main vehicle to convey the design concepts and ideas. However, my mentor Mr. Chen Congzhou once said: "I think that to study the Chinese garden, one should start with Chinese poetry in order to discover the essence from its source. Then it should be easier to resolve the issues related to it. If one is merely talking about gardens, one cannot fully comprehend them". How can a garden designer or the "seven-tenth-master" have design inspirations if he or she doesn't know about the Chinese aesthetics and doesn't have a good literary foundation? During the design process of Liu Fang Yuan, I tried to combine a designer and a scholar to create and convey

流芳园 － 太湖石、玉茗堂、映芳湖
LFY - The Taihu rockery, The Tea House, the Lake of Reflected Fragrance

带之情，山领回接之势，草木适掩映之容，亭榭点诗画之景，方能成园林之趣；造园，虽有奥思，仍需巧匠细工，方可得精致清秀之气象、诗情画意之境界。中国园林营造了一个“忘我”的境界，而西方园林创造了一个“有我”的境界。

每一座园林需要有它自己的“风月”！流芳园以其“九园十八景”为主题特征，结合汉庭顿的人文气息和自然环境，形成了她的独特园林气质和风格。虽然目前流芳园还没有完全建成，未来新增的部分会与总体规划有所差异，但是总体格局和主要内容不会有变化。当然，就总结经验来讲，在目前已经建成的部分当中也存在着一些由于材料、施工工艺、

the concepts and ideas through not only design drawings but also in verses and names for the scenes in the garden. In other words, a method of “literary thinking” has been involved in the “creative design” process. Every garden should present its own “painted” scenes and its own “described” sentiments. The Chinese garden design and construction creates scenes of mountains with magnificent and vigorous momentum, rocks with the spirit of slenderness, pierced, openness and wrinkles, meandering waters with beauty, flowers with symbolic implications for human virtues, and trees that are both quaint and age-old. All these start from setting foundations, arranging water courses, piling up hillocks, selecting rocks, and choosing trees and flowers. They all end up with “enthralling” the wind, the moon and the visitor as well.

流芳园 - 玉镜台内景

LFY - The inside of Terrace of the Jade Mirror

当地建筑和抗震规范等影响所产生的问题。比如，建筑方面有梁柱和栏杆的尺寸比例比中国传统建筑的要略粗大一些，因为需要加强结构的强度以满足当地的抗震要求；为方便残疾人游览而设的“无障碍通道”使得一些原本应该有台阶的地方被取消；以及当地气候环境对部分中国品种植物的生长和审美造成不尽理想等问题。因此，中国传统式园林若要适合当代人的生活方式还需要做适当的调整，这也是未来建造中国传统式园林的挑战之一。

Arts are created by human beings. Chinese art focuses on the importance of reaching the heavenly or natural realm through human artifacts, which through understanding nature, being part of nature and re-creating nature has the goal of reaching the integration and unity of humans with nature. Therefore, the orthodox principle in the art of Chinese garden is that "though man-made, it should look like it was something naturally created." The state of naturalness is the ultimate source of inspiration for garden-making. After designing Liu Fang Yuan and seeing its realization and evolution, my understanding and perspective on the art of Chinese garden has been enriched and renewed. I believe that the Chinese garden design and construction is a work of art which excels nature, fitting in with the terrain and applying manpower; it resembles natural imagery as well as imagery of the mind. It must make the waters with twining emotions, mountains tortuous, plants beautiful, pavilions ornamenting the poetic and pictorial scenery; all these will make the garden more vivid and attractive. Although the profound design concepts are well developed, skillful craftsmanship is essential to make the garden delicate and elegant, poetic and picturesque. The Chinese garden is a world where people forget themselves inside the garden, while the Western garden a state where people are aware of themselves.

Every garden needs to have its own *feng yue* or "wind and moon" or distinguished features. Liu Fang Yuan, with its thematic scenery of the "Nine Gardens and Eighteen Views", and within the Huntington's exquisite literary atmosphere and natural environment, has its one qualities and style. Although it is not completed as a whole, and the future phases may be differ from the master plan, the overall layout and the key components of Liu Fang Yuan will be the same as the master plan shown in this book. Certainly there are some lessons to be learned from the completed sections of the garden. For instance, there are some problems from the application of the local code and seismic requirements which caused the sizes of structures such as posts and beams to be bigger than traditional Chinese buildings, and railing heights are higher and thicker than the ones in Suzhou gardens. Liu Fang Yuan has been made with handicapped accessible routes through the main courtyards and buildings, but some steps have been eliminated, which loses some of the authenticity of traditional walkways. In the planting areas, because of the local climate, some plants grow too fast so that it is difficult to keep them in right shapes and sizes, etc. Therefore, it is one of the challenges to maintain the authenticity of the traditional Chinese garden, while make it serve modern society well.

致谢
ACKNOWLEDGEMENTS

“造园如作诗文”，而书写园记也犹如一次“造园”的经历，一次畅游园林的体验。在本书的写作过程中，得到了各方的大力支持和帮助。

首先，我要特别感谢汉庭顿总裁史蒂夫·郭必列先生多年来对我工作的信任、坚定的支持和巨大的鼓励。我还要感谢副总裁莱恩迪·肖曼以及前副总裁苏珊·莫尔女士所给予的支持和在汉庭顿档案资料方面给予的特别安排。没有他们的支持和帮助，本书很难呈现得完整。

我最深切的谢意是给汉庭顿植物园主任詹姆斯·富尔森先生的。感谢他对我的信任以及给予我设计流芳园的机会。在汉庭顿的三年工作时间里，我非常荣幸地能在富尔森主任的领导下工作，使我受益匪浅。

To make a garden is like to write a poem or literary work. To write a record for the garden is like an experience of making a garden and strolling in the garden. During the process of writing this book, I have personally benefited from the generous supports of the many.

First and foremost, I want to thank Steve Koblik, President of the Huntington, for believing in me and giving me his steadfast support and great encouragement over the years.

I am particularly indebted to Vice President Randy Shulman and former Associate Vice President Suzy Moser, for having agreed to provide the Huntington's achieved materials for this book. Without their support, this book would not be appeared as it is now.

My deepest gratitude goes to James Folsom, Director of the Botanical Gardens at the Huntington, for believing in me and giving me the opportunity to work on this marvelous garden project. I had also benefited from the privilege of working with him during the three years of my stay at the Huntington.

I am very grateful to Professor Zhu Liangzhi of Beijing University for spending his precious time reading the manuscript of this book and writing the preface for it. As a scholar in Chinese aesthetics and a great Chinese garden connoisseur after Chen Congzhou, Prof. Zhu has written many books and articles on the aesthetics of traditional Chinese gardens. His magnificent writings have inspired me on some of the conceptions in this book.

As a dear friend and partner, I want to thank Kwang-I Yu for his inspirational and invaluable

流芳园 － 景窗
LFY - garden window

我要特别感谢北京大学的朱良志教授在百忙当中阅读了本书文稿，并为本书作序。作为继陈从周先生之后，又一位中国园林美学和鉴赏大家，朱先生对中国古典园林美学的研究和阐述之深入和全面，给本书的写作以极大的启示。

我还要特别感谢我的好友俞匡一先生，他对本人这些年来从事的设计工作以及写作这本书都给予了宝贵支持和鼓励。他也是一位流芳园的赞助者。

我在洛杉矶生活和在汉庭顿工作期间，有幸与居住在洛杉矶地区陈从周先生的女儿陈胜吾和她先生陈平老师相识。多年来我们经常畅叙陈从周先生做人做事的风格及其对中国园林艺术的思想，真是一段十分愉快的经历，在此特别对二位表示感谢。

另外，本书中所使用的汉庭顿历史图片共计 13 张，全部是由汉庭顿图书馆提供的，对此我深表谢意。本书中所使用的园林照片主要由我本人拍摄而成，包括苏州和扬州园林和汉庭顿花园以及流芳园的照片共计 110 张。另外，朱宝康先生和刘华杰先生分别提供了 19 张和 16 张流芳园的彩色照片，这些照片给本书增色不少，对此我要向他们表示感谢。

最后，我要非常感谢当年与我一同工作在流芳园项目上的汉庭顿同仁、中美双方的设计和施工人员、当地社区的友好人士以及流芳园的赞助者和支持者们，是他们的热情、坚守和通力合作，使流芳园成为献给大众的一份优美的文化礼物。

support on my work in designs over years and the writing of this book. Yu is also a sponsor of Liu Fang Yuan.

When I was living in Los Angeles and working at the Huntington, I had pleasure to meet with Chen Shengwu, a daughter of Chen Congzhou, and her husband Chen Ping, who also lived in Los Angeles area. We had many conversations about Chen Congzhou's way of doing things as well as his ideology on Chinese gardens. It was a quite pleasant experience. I reserve a special appreciation for Chen Shengwu and Chen Ping.

I want to thank the Huntington Library for providing 13 images of the Huntington's historical materials. Most of the garden photographs used in this book are taken by the author, with a total of 110 photographs including the ones of Suzhou and Yangzhou gardens as well as Liu Fang Yuan. In addition, Kenny Chu has provided 20 pictures and Ricky Lau 16 pictures of Liu Fang Yuan. Those photographs have enriched the presentation of this book, for which, I deeply appreciated their support.

Finally, I want to thank my colleagues at the Huntington whom I had worked with on the Liu Fang Yuan project, people in the design and construction teams of this project from both China and the US, and the steady supporters from the local communities as well as the sponsors of the project. Their enthusiasm, persistence and collaborations have made the great garden of Liu Fang Yuan a truly magnificent cultural gift to the public.

结束语
AFTERWORD

中国园林是一个充满活力的奇妙世界。流芳园中的花香、流水和山石，会在岁月变化和四季轮回中呈现出不一样的魅力和不一样的气息；流芳园中的亭台廊榭，会因不同的游人在动观与静观之间而具有不一样的内涵；流芳园中蕴含的诗情画意，也会使不同的游人产生不一样的感受和意境。

陈从周先生在《说园》中说过："造园之学，主其事者须自出己见，以坚定之立意，出婉转之构思。成者誉之，败者贬之。无我之园，即无生命之园。" 通过对汉庭顿流芳园的设计，我对先生的观点体会得更深，也更有同感。在我 1983 年从同济大学毕业时，从周先生赠予了我一本《说园》，并在扉页上以我的姓名作了一幅嵌字对："疾风知劲草，陈陈莫相因"。他勉励我，做事业须坚忍不拔；搞设计要敢于创新，切莫因袭旧套。

The Chinese garden is an intriguing and vivid cosmos. Inside Liu Fang Yuan, the flowers, streams and rockeries will appear to have different charms and flavors during different times and seasons; the pavilions and walkways will have different connotations to different visitors when they look at them in-motion or in-still; and the poetic and picturesque sceneries will evoke different feelings and imaginations to different visitors as well.

Chen Congzhou said in *On Chinese Gardens*: "To design gardens, the designer ought to proceed from a conception peculiarly of his own and work out the layout of the garden adroitly without the slightest deviation from his original conception. A garden successfully built in this manner will be highly prized whereas a failure will surely induce criticisms. Success or failure, a garden which fails to embody the designer's personality is one devoid of life". After designing Liu Fang Yuan, I have the same feeling and agreed even more with my mentor's point of view. In 1983, right before I graduated from Tongji University, Professor Chen Congzhou gave me a copy of the first edition of his book *On Chinese Gardens* as a gift. He wrote a couplet which was composed from my name on the title page says: "The force of the wind tests the strength of the grass; persisting not in the old ways without any change." Then he encouraged me to be steadfast in my faith and be creative in designs.

As the chief designer of Liu Fang Yuan, after a keen observation of the site of the garden,

流芳园 － 海棠花门洞
LFY - the begonia-shaped gate

作为流芳园的总设计师，在对园址进行了敏锐和细致的勘察之后，我综合性地构思出流芳园的总体布局和园景设计，将其与汉庭顿的环境和场地的特质紧密结合。同时，在传承中国传统园林艺术和造园手法的基础上，我在设计创新方面进行了探索，比如将个别中国北方皇家园林的园林意境融合到流芳园的设计中，又比如在这座“文人园”中布置了一座宝塔等。似故而实新，似同而实异。这些创新的目的是让流芳园更好地反映出汉庭顿的总体文脉特征，更好地融入到南加州的地理气候环境当中，使流芳园在按照总体设计完成之后，能够成为一座具有独特的园林气质、令人陶醉的美景和深厚的文化内涵的中国园林。

380年前，明代计成完成了中国第一部造园专著《园冶》。古人说：“文以载道”。《园冶》讲述了明代时期中国江南造园的思想理论和法则，更讲述了中国的造园之“道”。优秀文化传统需要传承，中国园林文化和艺术的传承要靠保护好留存下来的园林实例，同时也要靠文字予以记录、总结和提高。本书是对一座新的尤其是在海外的中国传统式园林的创作过程、设计理念和成果展示的记录，希望它对理解和欣赏中国园林及其造园思想

I conceptualized and designed the master plan and scenery peculiarly in response to the unique characters of the site and the Huntington environment. In addition, with a reverence for the orthodox principles and the paradigm of Chinese gardens, I tried to create something new here, such as the reference to some poetic conceptions in the northern imperial gardens of China and the concept for a pagoda inside the so called “scholar garden”. What I have designed for Liu Fang Yuan is new even though it is reminiscent of the classical gardens in China. All these choices are aimed to make the garden better fit for the context of the Huntington as well as to the bigger geological and climatic environment of Southern California, which will make Liu Fang Yuan, once it is consummated as planned, a Chinese garden with distinctive qualities, enchanting beauties and profound cultural connotations.

Three hundred and eighty years ago, Ji Cheng, a garden master of Ming dynasty, finished his book of *Yuan Ye* or the *Craft of Gardens* which is the first monograph on traditional Chinese gardens. The ancient Chinese said: “Writings are for conveying the Dao (or the Way or truth).” *Yuan Ye* narrated the theory and concepts of Chinese gardens in Jiangnan region in Ming period, which in fact told the Dao of Chinese garden design and construction. Excellent cultural heritages need to be inherited. The culture and art of the Chinese garden need to be inherited by preserving the existing traditional gardens, and at the same time, by writing and recording the history of the gardens. This book of *Liu Fang Yuan* is a record of a new traditional style Chinese garden built outside of

和园景之外的文化内涵有所帮助。计成在《园冶》最后的"自识"中写道："崇祯甲戌岁，予年五十有三，历尽风尘，业游已倦，少有林下风趣，逃名丘壑中，久资林园，似与世故觉远……"本人著此《流芳园》一书之年亦正五十有三，虽业游已久，却尚未厌倦；虽常有林泉意趣，却仍怀卧游之心；虽仍有事业激情，却羞于才疏学浅。故今梓行，合为世鉴，雅正为盼。

冶园道人 陈劲

2014 年冬月　于海上知然居

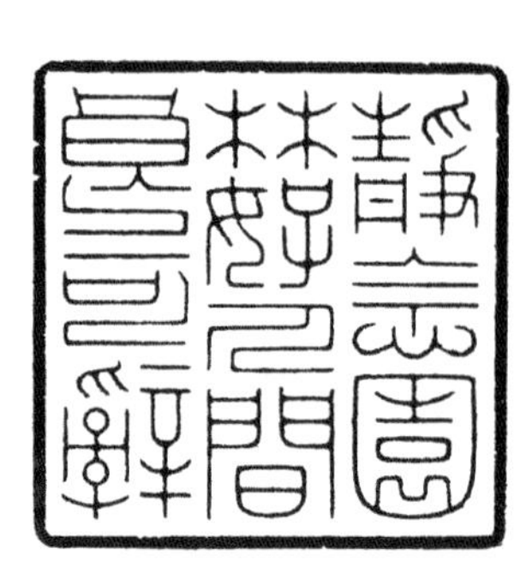

China, which tells the project background, the design process and ideas as well as photographic presentation of the finished sections of the garden. I hope that this book will be helpful to the readers to understand and appreciate the aesthetics and cultural connotations of the Chinese garden. In the "Afterword" of *Yuan Ye*, Ji Cheng said: "In the Jiaxu year of the Chongzhen era (AD 1634), I am fifty-three years old. I have gone through many rough hardships, and am already tired of my wandering in the course of my profession. I still have a little interest in woodland and evading fame among the hills and ravines. For a long time I have made a living from garden designs and constructions, while I have felt as though I was cut off from the things of this world……" This year when I am writing this book of *Liu Fang Yuan*, I am fifty-three years old too. Although having been in the design profession for a long time, I am not tired of doing this. I prefer to stay at home to read landscape paintings and dream of traveling in those landscapes instead, though I am occasionally fond of going to forests and streams. Although I am enthusiastic about my career, I am diffident due to my limited talent and knowledge. So herein, I have had this book printed and published for the discussion among the public; and I look forward to hearing your opinions.

Jin Chen

A Garden Daoist　Wrote in Zhi Ran Ju, Shanghai, China

December 2014

流芳园 — 玉镜台

LFY - Terrace of the Jade Mirror

主要参考文献
SELECTED BIBLIOGRAPHY

中文参考文献 References in Chinese

陈从周 Chen Congzhou 《说园》，同济大学出版社，2007 年。

《园林谈丛》，上海文化出版社，1980 年。

《园综》，同济大学出版社，2004 年。

《苏州园林》，上海人民出版社，2012 年。

冯友兰 Feng Youlan 《中国哲学简史》，北京大学出版社，2013 年。

姜义华 Jiang Yihua 《中华文化读本》，上海人民出版社，2009 年。

金学智 Jin Xuezhi 《中国园林美学》，中国建筑工业出版社，2005 年。

刘敦桢 Liu Dunzhen 《苏州古典园林》，中国建筑工业出版社，2005 年。

罗哲文 Luo Zhiwen 《中国古园林》，中国建筑工业出版社，2000 年。

潘谷西 Pan Guxi 《江南理景艺术》，东南大学出版社，2001 年。

邵　琦 Shao Qi 《中国画文脉》，上海书画出版社，2005 年。

童　寯 Tong Jin 《江南园林志》，中国建筑工业出版社，1987 年。

王启兴 Wang Qixing 《全唐诗》，湖北人民出版社，2001 年。

夏承焘 Xia Chengtao 《宋词鉴赏辞典》，上海辞书出版社，2013 年。

张家骥 Zhang Jiaji 《中国造园论》，山西人民出版社，2003 年。

《中国园林艺术大辞典》，山西教育出版社，2005 年。

张竞无 Zhang Jingwu 《陈从周讲园林》，湖南大学出版社，2009 年。

宗白华 Zong Baihua 《美学散步》，上海人民出版社，2014 年。

周积寅 Zhou Jiyin 《中国画论大辞典》，东南大学出版社，2011 年。

朱良志 Zhu Liangzhi 《生命清供》，北京大学出版社，2005 年。

《中国美学十五讲》，北京大学出版社，2006 年。

《中国艺术的生命精神》，安徽教育出版社，2006 年。

《扁舟一叶》，安徽教育出版社，2006 年。

《真水无香》，北京大学出版社，2009 年。

英文参考文献　References in English

Ji Cheng, *The Craft of Gardens.* Translated by Alison Hardie. Yale University Press, 1988.

The Huntington Library, Art Collections and Botanical Gardens, The Huntington Press, Second Edition 2003

Maxwell K. Hearn, *Cultivated Landscapes*——Chinese Paintings from the Collection of Marie-Helene and Guy Weill, the Metropolitan Museum of Art, Yale University Press, 2002.

作者介绍

陈劲，号劲草，别号冶园道人，美籍华人，著名中国园林建筑家和规划设计师。1961年出生于贵州省贵阳市，祖籍湖南省湘阴县。同济大学建筑系园林专业学士，师从著名中国园林艺术大师陈从周教授等；美国麻省大学景观设计硕士，师从 Nicholas Dines 教授（曾任美国哈佛大学景观设计系主任）等，美国俄勒冈大学建筑学硕士。美国加利福尼亚州注册景观建筑师，美国景观设计师协会会员，现居住在中国上海市。

陈劲先生已经从事园林景观和建筑规划设计三十余年。在中国和美国主创设计过多个中国传统园林和古城保护规划设计项目，包括 1982 年在陈从周先生指导下，对上海市“豫园”东部修复进行的第一次方案设计；1995 年至 2000 年间在美国俄勒冈州波特兰市，代表美方负责中国古典园林“兰苏园”项目的设计和施工协调管理；2001 年至 2003 年间在美国加利福尼亚州洛杉矶地区汉庭顿机构，负责主创了中国园林“流芳园”的总体规划设计；2005 年主创设计了苏州拙政园历史保护区及苏州博物馆地段的保护方案；2010 年主创设计了山东省青岛胶州市的古典园林“嘉树园”项目；2012 年至 2013 年间主创设计了山东省青岛即墨市的“即墨古城”复兴总体规划设计。曾接受过新华社和美国第三大报纸《洛杉矶时报》、美国最大中文报纸《世界日报》和《侨报》以及享誉好莱坞的《洛杉矶杂志》的专访。

About the Author

Jin Chen, aliases jin cao and a Garden Daoist, is an American Chinese, and a well-known Chinese garden architect, landscape architect and planner. Chen was born in the City of Guiyang, Guizhou Province in 1961, a native of the Xiangyin County in Hunan Province in China. He went to Tongji University to study landscape architecture as well as to learn Chinese garden design under the tutelage of Professor Chen Congzhou, a preeminent Chinese garden scholar in China. Jin Chen obtained his master’s degree in landscape architecture at the University of Massachusetts. His master thesis project was directed by Professor Nicholas T. Dines who was once the Director of Landscape Architecture Department at Harvard University; and then he obtained his master’s degree in architecture at the University of Oregon. Chen is a licensed landscape architect in the State of California in the United States, a member of American Institute of Landscape Architects. Jin Chen now lives in Shanghai, China.

Jin Chen has worked in the field of garden and landscape design for over thirty years. He has designed many gardens in traditional Chinese style and led historical preservation projects in many cities in both China and the US. Notable among these are the restoration design project of the Eastern Area of "Yu Yuan" directed by Chen Congzhou in 1982; between 1995 to 2000, Chen was the project coordinator and design consultant in Portland's Chinese Garden "Lan Su Yuan"; between 2000 to 2003, Chen was the chief designer and project manager in the Huntington's Chinese Garden "Liu Fang Yuan" project; in 2005, he led the project of the preservation plan for the District of the Humble Administrator's Garden and Suzhou Museum; in 2010, Chen designed the traditional Chinese garden of "Jia Shu Yuan" in Qingdao Jiaozhou city; and between 2012 to 2013, Chen led the restoration master planning project of "Jimo Ancient City" in Qingdao.

Jin Chen had been interviewed and reported by the *Los Angeles Times*, *Los Angeles Magazine*, *China Daily* and *Chinese Press* as well as the Xinhua News Agency.

流芳园 – 映芳湖

Liu Fang Yuan - Lake of Reflected Fragrance

图书在版编目（CIP）数据

美国流芳园设计：海外中国名园/（美）陈劲著.
—上海：上海人民出版社，2015
ISBN 978-7-208-12958-0

Ⅰ. ①美… Ⅱ. ①陈… Ⅲ. ①园林艺术—介绍—美国
Ⅳ. ①TU986.671.2

中国版本图书馆 CIP 数据核字（2015）第 090990 号

责任编辑 苏贻鸣 秦 堃
装帧设计 胡 斌

美国流芳园设计
——海外中国名园
［美］陈 劲 著
世纪出版集团
上海人民出版社出版
（200001 上海福建中路 193 号 www.ewen.co）
世纪出版集团发行中心发行
上海雅昌艺术印刷有限公司印刷
开本 635×965 1/8 印张 39 插页 4
2015 年 5 月第 1 版 2015 年 5 月第 1 次印刷
ISBN 978-7-208-12958-0/TU·6
定价 380.00 元